GUNNERS!

GUNNERS!
B-29 MACHINE GUNNERS IN THE KOREAN WAR

JAMES BLACKWELL

DEEDS PUBLISHING | ATLANTA

Copyright © 2025 — James Blackwell

The views expressed in this publication are those of the author and do not necessarily reflect the official policy or position of the Department of Defense or the U.S. government. The public release clearance of this publication by the Department of Defense does not imply Department of Defense endorsement or factual accuracy of the material.

ALL RIGHTS RESERVED—No part of this book may be reproduced in any form or by any electronic or mechanical means, including information storage and retrieval systems, without permission in writing from the authors, except by a reviewer who may quote brief passages in a review.

Published by Deeds Publishing in Athens, GA

www.deedspublishing.com

Printed in The United States of America

Cover and interior design by Deeds Publishing

ISBN 978-1-961505-46-9

Books are available in quantity for promotional or premium use.
For information, email info@deedspublishing.com.

First Edition, 2025

10 9 8 7 6 5 4 3 2 1

Contents

Foreword	*ix*
1. Gunners & Gunnery	1
2. Killers	11
3. Philip Aaronson	23
4. Dale H. Crist	94
5. Thomas W. Stevens	102
6. Romaine Gregg	121
7. Jack Bernaciak	141
8. Who Were These Men?	157
Afterword	*163*
Acknowledgments	*166*

Foreword

In the annals of military aviation, there are repeated occasions of remarkable courage and superb airmanship. But, an underappreciated community are those that perform in typically large, lumbering and sometimes vulnerable bombers. These are aircraft that can deliver punishing strikes, provided they make it to the target. And, they depend on multiple crewmembers to accomplish any mission.

Officers and enlisted airmen comprise such crews. The pilots, navigators, bombardiers and electronic warfare officers are the officers of the crew. Less well known, but no less essential, are the flight engineers, radio operators and, yes, the gunners. All must perform in a coordinated and purposeful manner to ensure survival and mission accomplishment. This book is about the gunners who defended their aircraft from attacking fighters, initially with only their own experience and spatial skills, then increasingly with assistance of early computers to optimize muzzle velocities, two-body aerial geometry and ammunition ballistics. Still, a successful defense was frequently the work of teenagers or airmen in their early twenties. When the enemy succeeded in their attack, bomber crews did not return to base and family or they were shot down and became prisoners of war.

"Gunners" is a riveting story of the mid-20th century bomber era and those that performed the mission. Some of these exceptional airmen faced dire circumstances in captivity during the Korean War. Their experience, survival and resistance in awful conditions serves as testimony to their individual endurance and to the place they occupy as iconic American Airmen. Read on.

General Norton A. 'Norty' Schwartz, USAF, Retired
former Chief of Staff, United States Air Force

CHAPTER ONE

Gunners & Gunnery

Staff Sergeant Philip Aaronson delayed opening his parachute for 10 seconds rather than the standard four-count. The gunner wanted to make sure he was clear of the doomed B-29. He could see 10 other parachutes above him. He looked below and saw a solitary human shape plummeting to the ground. He looked up again and saw the B-29 spiraling through the air.

Suddenly, a thunderous, reverberating roar indicated the engine fire must have reached the fuel tanks and napalm bombs still in the bomb bay. The gigantic airplane instantly shattered into bits of metal and debris and was gone. Philip turned his attention to the earth rushing up towards him. He hit the ground, absorbing the shock exactly as he had been trained. Nothing was injured. He felt lucky. Then, North Korean soldiers arrived and fired shots over his head to make sure he knew he was captured.

Technical Sergeant Dale Crist was part of a B-29 crew trained to deliver atomic bombs. The atomic bomb couldn't be armed until the aircraft was airborne, and, though he was a gunner, he had to go out into the bomb bay to arm the bomb. Once armed, it couldn't be disarmed. Though the training flights used dummy bombs, they did contain explosives. That training meant survival on one combat mission over North Korea using conventional bombs. Using a screwdriver and perfect synchronization with the bombardier, Dale manually released an armed bomb through the open bomb bay doors at 10,000 feet.

Staff Sergeant Thomas W. Stevens was part of a Korean War B-29 crew that literally flew on the cusp of World War III. The gunner recalled that as his bomber was completing its bombing run over North Korean military facilities along the Yalu River—a strip of land notoriously known as MiG Alley for the Soviet fighter jets defending the area—a strong wind blew the bomber across the Yalu River. They were thus flying through Chinese airspace, a nightmare scenario for President Harry S. Truman.

Sergeant Romaine Gregg, who was only 16 when he enlisted in the Air Force, saw from his tail gunner's perch a new type of weaponry unleashed on the American bombers on one mission through MiG Alley. It wasn't the orange-glow golf ball explosion of artillery shells, nor was it the red tracer rounds fired from a MiG-15. It was a yellow streak of flame emitting from the tail-end of a rocket, the only time B-29 crewmen reported enemy missile fire during the Korean War.

Jack Bernaciak was tail gunner aboard the last B-29 on the final combat mission of the Korean War. While the 307th Bombardment Wing mission number 573 was in flight, sometime after "bombs away," the negotiators at Panmunjom signed the Armistice Agreement. He was thus the "tail end Charlie" of the bombing campaign.

Epic *Masters of the Air* encounters between US bombers and German fighters vividly seat you inside a World War II B-17 in combat. It's scary real. You can see and feel how gunners had an almost impossible task to pick up, identify, track, target, and shoot attacking enemy fighters within about five seconds.

Movie star Clark Gable famously enlisted in Wowrld War II specifically to become a B-17 gunner. He flew at least five combat missions at the waist gunner position, receiving the Distinguished Flying Cross, Purple Heart, and Air Medal.[1]

---

1. American Air Museum in Britain, "William Clark Gable." https://www.americanairmuseum.com/archive/person/william-clark-gable

Somber poetry still haunts eighty years later:

From my mother's sleep I fell into the State,
And I hunched in its belly till my wet fur froze.
Six miles from earth, loosed from its dream of life,
I woke to black flak and the nightmare fighters.
When I died, they washed me out of the turret with a hose.

-The Death of the Ball Turret Gunner,
by Randall Jarrell, 1945[2]

B-29 gunners faced great challenges in the Korean War. The big bomber designated as the Superfortress served in the last 14 months of World War II. It was thus the U.S. Air Force's primary bomber when the Korean War started in 1950. Soviet fighters were faster and more maneuverable than any German or Japanese World War II fighter. Or American. The B-29's remote-controlled turret machine guns were not designed to match Russia's nearly supersonic jets of 1950.

The men behind those machine guns in the Korean War were no different than those of the Greatest Generation. But Korean War gunners had no poetic figures or movie stars among their ranks. They were ordinary men who gave extraordinary effort.[3]

B-29 gunnery was radically different than gunnery on any previous American bomber. No more standing behind the gun, traversing and elevating, pressing the butterfly trigger and walking tracers into the target. The B-29 was outfitted with machine guns mounted in five

---

2. The Poetry Foundation, https://www.poetryfoundation.org/poems/57860/the-death-of-the-ball-turret-gunner
3. Nicholas Hobbs, ed. *Psychological Research on Flexible Gunnery Training* Army Air Forces Aviation Psychology Program Research Reports, Report No. 11, 1947, 268-270.

remote control turrets: two on top of the fuselage, two on the underside, and one in the tail. Each turret housed two or more M3 calibre .50 machine guns, an adaptation of the trusty Browning M2, modified for remote firing.[4]

Five gunners sat at control stations inside the B-29. The central fire control gunner (CFC) perched at the top of the fuselage under a bullet-proof plexiglass "blister" just above the aluminum surface of the aircraft. It gave him a 360-degree view of oncoming threats and friendlies. Below and to the sides of the aircraft sat two side gunners, called "scanners." These men could see beside and below the aircraft. Their duties also included serving as eyes for the men on the flight deck, observing the engines, propellers, and control surfaces on the wings. In the forward section, the bombardier doubled as a gunner with his own machine gun controls next to his Norden Bombsight. The best view of all was the tail gunner's, whose panoramic windows to the rear and sides of the bomber made him the protector of the 6 o'clock view.

Each station had a seat for the gunner, a sight with trigger buttons, and a control box with the switches and circuit breakers to control the functions of the M3 machine guns. Each gun turret had two to four of these trustworthy calibre .50 weapons. The sight was connected to the guns electrically through computers that instantly solved the complex math the gunner needed to get bullets on the target. Each gunner could control some of the five turrets, all coordinated by the central fire control gunner.[5]

B-29 Gunnery School was at Lowry Air Force Base near Denver, Colorado. The course took about six months compared to World

---

4. "Browning Aircraft Machine Guns Part III – High Speed and the Modern .50 Caliber Aircraft Guns," *Small Arms Review*, September 1, 2002. https://smallarmsreview.com/browning-aircraft-machine-guns-part-iii-high-speed-and-the-modern-50-caliber-aircraft-guns/
5. "Why the B-29's Gun System was so Effective," August 25, 2022 https://youtu.be/vwNPJgNEyMU

War II's six-week gunnery school. The B-29 Remote Control Turret System was complex, and gunnery training began with a turret course that ran about two weeks, concentrating on basic AC and DC electricity followed by troubleshooting circuits before the students even sat down at a machine gun. This was followed by instruction on how each of the bomber's five turrets worked and the complicated, computerized sighting system.

To accommodate large numbers of trainees, classes were conducted in three shifts. After reveille at 5 am, first-shift trainees rushed through their daily shave, uniform and barracks inspection, and breakfast routine, then marched across the active runway to class. Their corporal had a handheld radio for contact with the control tower to ensure safe crossing amid the constant flow of aircraft take-offs and landings at the busy base. They returned from class in similar fashion.[6]

Several weeks of aerial gunnery followed the turret course. They did not have to break down and re-assemble the machine gun with their eyes closed, as told in Army and Marine Corps mythology. But Air Force students did have to master the basic functioning of the venerable machine gun invented by John Browning after World War I. The men developed proficiency in tracking and aiming in electro-mechanical simulators, then flew daily for practice in the skies aboard an operational B-29.

Flying was what it was all about, and the men went through several exercises to prepare them for the high-altitude mission of the B-29. They learned how to put on oxygen masks and breathe the life-sustaining gas during the bomb run when the cabin would be depressurized in case the fuselage was penetrated by anti-aircraft artillery shrapnel or enemy fighters' bullets. Gunners also had to regulate the oxygen to the correct pressure flowing from the tanks.

---

6. Thomas W. Stevens, Korean War Legacy Project, 2024. https://koreanwarlegacy.org/interviews/thomas-w-tom-stevens/

Outdoors, gunnery students practiced bail-out procedures with hands-on training in how to exit the aircraft, deploy their parachute, guide the canopy to a safe spot on the ground, and execute a proper parachute landing fall.

The highlight of flight orientation week was the altitude chamber, where the pressure was reduced to that of around 25,000 feet. The men wore oxygen masks, but when they least expected it, the instructor ordered them to remove the masks so they could begin to experience hypoxia as the brain is starved of oxygen. With masks off in the low-oxygen environment, trainees had to do simple arithmetic problems and write their name before re-donning their masks. Many did not recognize how much capability they had lost until, after the exercise ended and they had recovered, they could see for themselves how impaired they had become after reaching the end of the time of useful consciousness.

They also practiced performing duties as scanners: the eyes and ears for the pilots and flight engineer, checking to see that the flaps were down for takeoff and the gear was up and inside the fuselage once they were airborne. In flight, the gunners scanned the skies to alert the crew of approaching aircraft, friendly and enemy.

Once qualified to fly, the students flew on training missions from Lowry to a gunnery range in South Dakota to fire at targets on the ground. The exercises progressed to engagements with American fighter aircraft that would fly typical attack patterns on the B-29s. Gunners practiced the hand-eye coordination challenges of keeping their sights on a moving target. When they pressed the trigger buttons, cameras recorded the sight picture to score the student's proficiency.

On successfully completing the course, students were awarded their gunner's wings and assigned to combat crews. They transferred to Randolph Air Force Base to develop the functional and crew skills necessary to fight as a team.

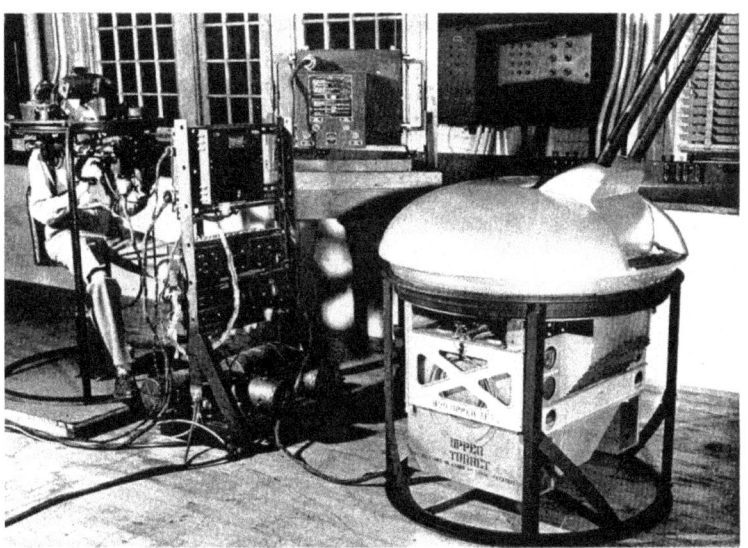

AAF Sergeant is shown operating basic elements of the B-29 central gunnery control system. All the gunner has to do is get the enemy plane in his sights and push the trigger button. The complicated mathematical problem of accounting for speed of the enemy plane, its distance from the B-29, gravity and parallax are worked out at split second tempo by electronic and mechanical units of the system. Smithsonian Institution, National Air and Space Museum, Industrial Aviation Magazine, vol. 2, no. 3, March, 1945. (https://airandspace.si.edu/multimedia-gallery/b29ia4503armamentcontrolp026wpng)

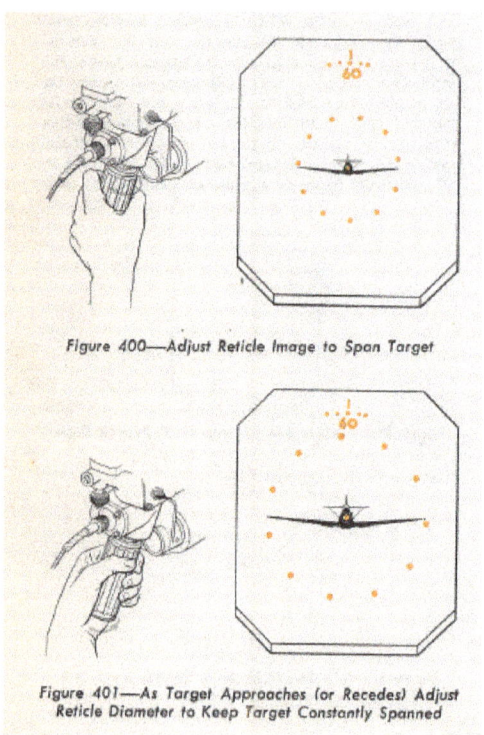

Figure 400—Adjust Reticle Image to Span Target

Figure 401—As Target Approaches (or Recedes) Adjust Reticle Diameter to Keep Target Constantly Spanned

World War II gunners sighted enemy aircraft by lining up iron sights along the barrel of the machine gun on the target then using their marksmanship skills to adjust the strike of the bullets to a point where the target was going to be when the bullet reached it. In the B-29 the gunner kept the target centered in a glass sight with an adjustable electronic reticle. The gunner needed to know the wingspan of the enemy aircraft and dial it in. When the target came into view the gunner used his hand controls to keep the target centered inside the sight reticle. Computers tracked the position of the sight controls to instantly calculate the exact point of aim then sent electrical signals to servo motors that rapidly and smoothly moved the turrets so that the gun barrels pointed at the precise spot in the sky to score multiple hits. (United States Air Force, Handbook, Operation and Service Instructions: Central Station Fire Control System T.O. No. 11-70AA-9 (Washington, DC: Authority of the Secretary of the Air Force, 27 March 1951), 178)

One week of combat crew training at Randolph Air Force Base as scheduled for a B-29 crew prior to deployment to the Korean War. These are unique skills to the B-29 bomber and must be mastered by the gunners assigned to each crew. Note the Gunnery Team Drills on Saturday and the special topics taught in the classroom such as "Psychological Warfare" and "General Effects of Atomic Bomb." (Author's personal collection)

Gunner's wings awarded on successful completion of aerial gunnery course in World II and the Korean War. TSgt. Harold L. Craven, USAF, "The Mark of and Eagle," Air Force Magazine, October 1963, 45.

CHAPTER TWO

Killers

Combat is about killing people and breaking things. This is an almost unspeakable truth. Many combat veterans have a story to tell about the enemy he or she killed. Most would rather not recall it but cannot erase the image.

A B-29 gunner would be in Korea for six to nine months. On average he would have about twenty-five combat missions, roughly one every nine days. On a twelve-hour mission, the aircraft would be under attack by ground fire during its bomb run, about five minutes, while an attack by a MiG-15 would last about 10 seconds. Korean War B-29 gunners, like infantry soldiers with combat experience, tell us that in those moments you fight like you are trained. Nothing else matters but killing the enemy who is working his hardest to kill you.

B-29 Gunners were killers. 22 Superfortress gunners shot down 25 MiG-15s in the Korean War. That's the record of officially credited enemy aircraft destruction. Three of those gunners destroyed two MiGs each, and one B-29 scored five MiG-15 kills—a bomber ace—with different crews.[7] Post-mission interrogations recorded many more but were not subsequently verified in formal reviews of wartime official records.

MiG-15 fighter pilots shot down 16 B-29s over North Korea.[8]

---

7. "Far East Air Forces Official Credit for Destruction of Aircraft 1951-1953," Air Force Historical Research Agency, Reel K1108, 396-670.
8. Korean War Air Loss Database (KORWALD), March 11, 2015 https://www.

On what Russian analysts critically misnamed "Black Thursday," April 12, 1951,[9] Soviet MiGs took down two B-29s from the 19th Wing. And, in an awfully branded account by an American memoirist on "Black Tuesday," October 23, 1951,[10] the 307th lost three to MiG-15s. These mischaracterizations seem to have carried forward among other publications on the B-29 in the Korean War.

But these B-29 gunners showed no such despair. Despite their losses, these fearsome men made the communists pay a high price, shooting down a total of 15 MiG-15s in those two air battles. The 307th Bombardment Wing's gunners took out a total of five MiG-15s in the air over North Korea during the war.[11]

When the Russians came after the B-29s again four days after the October 23 engagement, B-29 gunners wrought a leaded revenge, shooting down five more MiG-15s without loss of a single B-29.[12] Though any loss of life is tragic, the B-29 achieved a more than one-and-a-half-to-one superior kill ratio.

Official narratives portray the stuff that B-29 gunners were made of in the Korean War.

Richard J. Fisher

On November 14, 1950, just four days after a Soviet MiG shot

---

koreanwar.org/dpaa/korwald-all.pdf
9. Igor Seidov, *Red Devils Over the 38th Parallel: A Chronicle of Soviet Aerial Operations in the Korean War 1950-53*, West Midlands, UK: Helion & Company Limited, 2014, 123.
10. Earl J. McGill, Lt. Col. USAF (RET), *Black Tuesday Over Namsi: B-29s vs MIGs – The Forgotten Air Battle of the Korean War, 23 October 1951* (West Midlands, UK: Helion & Company Limited, 2011),141.
11. "Far East Air Forces Official Credit for Destruction of Aircraft 1951-1953," Air Force Historical Research Agency, Reel K1108, 396-670.
12. Robert A. Mann, *The B-29 Superfortress Chronology 1934-1960*, Jefferson, NC: McFarland & Company., Inc., 2009, 230. "Far East Air Forces Official Credit for Destruction of Aircraft 1951-1953," Air Force Historical Research Agency, Reel K1108, 396-670.

down the first B-29 in the war, the 307th Bombardment Group's target for mission number 56 was a bridge on the Yalu River at Sinuiju. This time they changed the ordnance load from incendiaries to 1,000-pound bombs.

The pre-mission blessing was held in the open: Typhoon Clara had blown away the large briefing tent three days earlier. Twelve aircraft from the 307th executed this mission. Flak was again intense, damaging three B-29s and wounding three crew members. Just after "bombs away," enemy MiGs attacked.[13]

This time the Soviet MiGs scored zero B-29 kills while the 307th killed one and damaged two others.

> ***Staff Sergeant Richard J. Fisher:*** *As the enemy came in for his attack, the tail gunner opened fire but was unable to continue because of the limiting cone of his fire. The enemy fighter came within the cone of the upper and aft forward turret which were controlled by Sergeant Fisher. Sergeant Fisher continued firing burst after burst into the fuselage and wing just forward of the bubble canopy. The enemy aircraft continued emitting heavy smoke and crashed into the ground.[14]*

David R. Stime and Ercel S. Dye

For Mission Number 151 on April 12, 1951, Bomber Command tasked all three B-29 groups to strike the north railroad bridge over the Yalu River at Sinuiju. At the initial point, on final approach to the target, the formation encountered a 125-knot crosswind that pilots struggled to keep from crossing the river into China yet still ensure the bombs would drop to the south on the bridge. They flew right by the Antung airbase, home to dozens of Soviet MiGs that took to the

---

13. 307th Bombardment Wing History, October 1950, AFHRA Reel N064, 1291.
14. "Far East Air Forces Official Credit for Destruction of Aircraft 1951-1953," Air Force Historical Research Agency, Reel K1108, 396-670.

air to intercept the Americans who had just violated the U.S. self-imposed "no fly" line.

Two gunners of the 307th each knocked off a MiG-15 that day.

Staff Sergeant David R. Stime *officially credited with the destruction of an enemy MiG-15 type aircraft in aerial combat near Sinuiju, North Korea at 1029I on 12 April 1951. Sergeant Stime was left scanner in a B-29 flying in formation. Two MIG-15's came in at 7 o'clock low and opened fire at about 800 yards. One (1) enemy aircraft broke off at 400 yards while the other continued to press its attack to 300 yards before breaking away. Sergeant Stime started firing at both aircraft at about 1,000 yards and continued to fire as they pressed home their attack. One (1) enemy aircraft broke off at about 400 yards rolling over and breaking away to the right and up. Sergeant Stime continued to fire at the other aircraft and observed his fire to hit the enemy aircraft in the right wing with smoke and small pieces flying through the air. The aircraft seemed to fall off to the right and started down in a vertical dive after trying to regain control. Enemy aircraft passed from beneath the B-29. When last seen, enemy aircraft had smoke coming from the right wing and was in a vertical dive. Pilot of fighter escort aircraft followed this enemy aircraft down and observed it to crash in the Sinuiju area.*[15]

Staff Sergeant Ercel S. Dye *is officially credited with the destruction of an enemy MiG-15-type aircraft in aerial combat near Sinuiju, North Korea, at 1026I on 12 April 1951. Sergeant Dye was tail gunner in a B-29 flying in formation. A MiG-15 was observed to turn in on the B-29 from 2,000 yards high, firing from 6*

---

15. "Far East Air Forces Official Credit for Destruction of Aircraft 1951-1953," Air Force Historical Research Agency, Reel K1108, 396-670.

o'clock, with slow overtake coming in level with the B-29. Sergeant Dye started firing at 1,000 yards and fired continuously down to 50 feet at which point the enemy aircraft broke off into a sharp descent trying to level off. Unable to level off, the enemy aircraft was observed going immediately into a vertical dive. A white smoke was observed to be coming from the aircraft as it headed toward the ground. Aircraft was seen to crash and explode on the ground.[16]

Three days later, on April 15, the 307th celebrated its 150th mission of the Korean War with an elaborate party in the Kadena Mess Hall with music and decorations. The Wing Commander sliced a huge cake, all Airmen enjoyed a steak dinner, and recognized Staff Sergeants Stime and Dye for their lethal accomplishments.

Jerry M. Webb and Fred R. Spivey

Tuesday, October 23, 1951, will long be remembered by members of the 307th Bombardment Wing. On this day, nine aircraft took off to strike Namsi Airfield, North Korea. Three were shot down over the target within a matter of minutes. Five more aircraft suffered extensive damage, with only one escaping completely undamaged.[17]

But a stricken Long Ranger (the men of the 307th Bombardment Group earned that moniker during World War II because of the extreme distances from their base to their targets) is a fearsome foe. Two gunners of the 307th scored MiG kills that day.

Staff Sergeant Fred R. Spivey is officially credited with the destruction of one MiG-15-type aircraft in aerial combat near Namsi, North Korea, at 0940I on 23 October 1951. Sergeant Spivey was a central fire control gunner on a B-29 that was part of a

---

16. "Far East Air Forces Official Credit for Destruction of Aircraft 1951-1953," Air Force Historical Research Agency, Reel K1108, 396-670.
17. 307th Bombardment Wing History, October 1950, AFHRA Reel N065, 192.

daylight formation attacking Namsi Airfield, North Korea. While on the initial point approaching the target, Sergeant Spivey's formation was attacked by an estimated 50 MIG-15 fighters. In the course of one of these attacks, a MiG coming in from one o'clock sustained heavy hits from a range of 1,200 yards until it was 10 yards away. After passing over this formation, the MiG was observed to be emitting smoke, with parts falling off. Continuing on downward, this MiG-15 was last seen falling out of control.[18]

Staff Sergeant Jerry R. Webb *is officially credited with the destruction of one MiG-15-type aircraft in aerial combat near Namsi, North Korea, at 0940I on 23 October 1951. Sergeant Webb was a tail gunner on a B-29 that was part of a daylight formation attacking Namsi Airfield, North Korea. During the course of the bombing run on Namsi Airfield, this formation of which Sergeant Webb's plane was a member, was subjected to intense and violent attack by an estimated 50 MiG-15 fighters. On one of those attacks, two MiGs came in from six o'clock level and within range of Sergeant Webb's guns. One of the MiGs was observed to receive a great number of hits at a point near the wing root, causing it to swerve away, smoking. Further enemy contact precluded observation, but other crew members observed this MiG to disintegrate in mid-air before it broke away.*[19]

Those FAGOT pilots ("FAGOT" was the official intelligence code name for the MiG-15 and had no relevance to the British or colloquial meaning of the term) did not live to tell their side of the story at the Soviet fighter bar that night.

---

18. "Far East Air Forces Official Credit for Destruction of Aircraft 1951-1953," Air Force Historical Research Agency, Reel K1108, 396-670.
19. "Far East Air Forces Official Credit for Destruction of Aircraft 1951-1953," Air Force Historical Research Agency, Reel K1108, 396-670.

Let no one, American or Russian, say that the 307th had a "Black Tuesday," or a "Black Thursday." They arose, flew, fought, and completed each assigned mission. They did it again the next mission. And the next, and the next, and the next, until war's end on July 23, 1953, on mission number 573.

In this photo, S/Sgt Fisher is the middle of the three airmen. Left to right are 1st Lt Leo Beinhorn, Navigator; Staff Sergeant Richard J. Fisher, side gunner; and S/Sgt C.E. Hall, tail gunner; in their post-mission interrogation. (Department of Defense Photo, National Archives, RG 342 Box 3049)

Staff Sergeant Ercel Dye boards his B-29 through the rear entrance door to make his way and take his position as Tail Gunner. (Photo: Department of Defense, National Archives, RG 342 Box 3059)

Staff Sergeant Jerry Webb points to the 37mm cannon direct hit behind him that knocked out the rudder controls, while another hit damaged the hydraulic system. (Photo: Department of Defense, National Archives RG 342 Box 3059)

In this photo, three of the 307th's gunners pose for their reconstruction of the October 23, 1951, air battle for the post mission intelligence section interrogation. Staff Sergeant Jerry Webb, on the left, holds a red model jet. Webb, the tail gunner on 44-70151, received official credit for killing a MiG-15. On the right, Technical Sergeant Malcom Fairchild, central fire control gunner, holds a model of a Superfort and whistles at their close call. He was initially credited with one destroyed and another probably damaged. In the center, Technical Sergeant Frank Bata, right gunner, lost his cap and headset when MiG shells ripped through the dome at his position, showering him with plexiglass. (Caption and Photo: Department of Defense: National Archives RG 342 Box 3059)

A burst of communist flak ripped large portions of skin surface from the tail end of a USAF B-29 Superfort while the 307th Bomb Group was on a daylight bombing attack of Red targets over North Korea. Other flak shrapnel made numerous punctures of the Superfort, but the huge FEAF Bomber Command plane returned its crew safely to an Okinawa base. The B-29 first demonstrated its ability to take terrific punishment during World War II and has maintained that reputation during the course of the Korean air war. January 1952. (Caption and Photo: Department of Defense: National Archives RG 342 Box 3059.)

USAF Official Credit for Destruction of Enemy Aircraft in Korean Conflict

What the Pundits Said	Date	MiGs Downed by B-29s	B-29s Downed by MiGs	What the Data Says
	10 November, 1950		1	
	14 November, 1950	1		
	30 March, 1951	2		
	7 April, 1951		1	
Black Thursday	12 April, 1951	6	2	Russian Widow Maker
	31 May, 1951	1		
	1 June, 1951	2	1	
	9 July, 1951	2		
	10 October, 1951	1		
	22 October, 1951		1	
Black Tuesday	23 October, 1951	3	3	Soviet Hell Week
	24 October, 1951	1	1	
	27 October, 1951	5		
	10 June, 1952		1	
	8 November, 1952		1	
	9 November, 1951	1		
	19 November, 1952		1	
	30 December, 1952		1	
	10 January, 1953		1	
	29 January, 1953		1	
B29 Driven out of the sky	February - July, 1953			
	TOTAL	**25**	**16**	

B-29 v MiG-15 Kill Ratio = 1.563:1

Overall US v Enemy Aircraft Kill Ratio = 1.007

Sources
Far East Air Forces Official Credit for Destruction of Aircraft 1951-1953, (Maxwell Air Force Base, AL: Air Force Historical Research Agency, Reel K1108), 396-670.
Korean War Project, Korean War Air Loss Database, KORWALD 15 March 2015 https://www.koreanwar.org/dpaa/korwald-all.pdf
Robert F. Futrell, The United States Air Force In Korea 1950-1953, Revised Edition, (Washington, D.C.,: Office of Air Force History United States Air Force, 1983), 692
FEAF Bomber Command (Provisional) History, (Maxwell Air Force Base, AL: Air Force Historical Research Agency, Reel K1108), pp. 376 - 651
"USAF Historical Study No. 81" USAF Historical Division Aerospace Studies Institute Air University June 1963

CHAPTER THREE

Philip Aaronson

"... they're going to kill you!"[20]

Staff Sergeant Philip Aaronson was central fire control gunner on the first B-29 shot down by a MiG-15 over North Korea. His experience as a gunner is unique. He fought communist forces in the air for five months and on the ground for almost three years.

Philip was born on September 5, 1925, in Miami, Florida. His parents, Hyman and Dorothy, raised him, his two brothers and five sisters in Harrisburg, Pennsylvania.[21] The Aaronson family were observant Jews who joined in celebration of America's patriotic and religious holidays. Philip grew into a gregarious fellow as he matured at Camp Curtin Academy Junior High School.

At William Penn High School, Philip played junior varsity football in the ninth grade, acted in school performances, and served as announcer for school programs.[22] In his senior year, December 1942, his sister Norma married Army Lieutenant Robert H. Cohn, who was in training to become a pilot. Cohn was a publicity manager for Columbia Studios in Hollywood before enlisting.[23] This wedding, with

---

20. Affidavit of Major David F. MacGhee, subscribed and sworn to by Major Frank M. Finn, JAGC, 13 April 1954, 2. http://www.kpows.com/amazingveterans.html
21. "SSGT Philip Aaronson," *The Patriot News*, Harrisburg, PA, March 31, 1998.
22. "Camp Curtin Notes," *The Evening News*, Harrisburg, PA: November 7, 1940, p. 29; "Curtin Class Holds Party," Harrisburg Telegraph, Harrisburg, PA June 5, 1941, 14.
23. "Will Become Bride of Lieut. Robert H. Cohn," *The Evening News*, Harrisburg,

military honors in Ohev Sholom Temple, had a compelling influence on Philip's motivation to serve in the Air Corps himself.

After graduation in 1943, Philip enlisted in the Army Air Forces.[24] During his time at various training camps and gunnery schools he was frequently attached to the entertainment division as master of ceremonies for camp shows.[25] His faith, personality, strength, and experience would serve him and his crewmates well in war.

Corporal Aaronson graduated from Flexible Gunnery School at Kingman, Arizona, in October 1944.[26] The Army assigned him as a B-17 ball turret gunner with the 96th Bombardment Group at Snetterton Air Base, England, in March 1945 where he completed pre-combat qualification in May.[27] He remained in England until redeployment back to the United States in April 1946.

Sergeant Aaronson reenlisted in May 1946 and was initially assigned to 2nd Fighter Squadron Headquarters, 52nd All Weather Fighter Group, Mitchel Air Base, NY.[28] He was subsequently assigned to the 30th Squadron, 19th Bombardment Group at North Field, Guam.[29] The Gunnery School at Lowry Air Force Base was not

---

PA, December 8, 1942, 6.
24. Aaronson, Philip, Application for Word War II Compensation, May 18, 1950. https://www.ancestry.com/discoveryui-content/view/347252:3147?tid=&pid=&queryId=318211ba-920d-49d5-96f2-b31dc7c71164&_phsrc=FEh95&_phstart=successSource
25. Aaronson, Philip, Application for Word War II Compensation, May 18, 1950. https://www.ancestry.com/discoveryui-content/view/347252:3147?tid=&pid=&queryId=318211ba-920d-49d5-96f2-b31dc7c71164&_phsrc=FEh95&_phstart=successSource
26. "Gunnery School Graduate," *The Evening News*, Harrisburg, PA, October 24, 1944, 7.
27. "Trains in England," *The Evening News*, Harrisburg, PA, May 9, 1945, 24.
28. *"Sent to Mitchel Base," The Evening News, Harrisburg, PA, July 29, 1948, 10.*
29. Aaronson, Philip, Application for Word War II Compensation, May 18, 1950. https://www.ancestry.com/discoveryui-content/view/347252:3147?tid=&pid=&queryId=318211ba-920d-49d5-96f2-b31dc7c71164&_phsrc=FEh95&_phstart=successSource

yet operational, so he transitioned to B-29 gunnery through on-the-job training.³⁰

The day after the Korean War broke out in June 1950, the 19th Bombardment Group started moving to Kadena Air Base, Okinawa.³¹ From there, B-29s conducted daily strikes, initially on advancing North Korean forces along the Han River near Seoul,³² then on enemy lines of communication farther north.³³ By August, B-29s were committed to providing close support to United Nations forces defending against North Korean troops threatening to push U.S. forces off the Pusan perimeter and into the Sea of Japan.³⁴

Philip flew on just about every mission assigned to the 19th Bombardment Group in the first weeks of the Korean War, but he had few opportunities to lead his gunners in defending his B-29 from attacking enemy fighters. Fighter attacks on B-29s at this stage in the Korean War were few and far between.

The North Korean Air Force was small. Though their pilots and their aircraft—the Soviet propeller-driven Yak-9—were capable World War II veterans, 18 of the 70 or so in the North Korean force were shot down by U.S. aircraft, and uncounted more were destroyed on the ground.³⁵ They rarely dared to take on the American B-29,

---

30. 19th Bombardment Wing History, October 1950, AFHRA Reel B0921, 1211-1579.
31. Robert F. Futrell, *The United States Air Force in Korea 1950-1953*, Revised Edition, (Washington, DC: Office of Air Force History United States Air Force, 1983), 24.
32. Robert F. Futrell, *The United States Air Force in Korea 1950-1953*, Revised Edition, (Washington, DC: Office of Air Force History United States Air Force, 1983), 29.
33. Robert F. Futrell, *The United States Air Force in Korea 1950-1953*, Revised Edition, (Washington, DC: Office of Air Force History United States Air Force, 1983), 24-25.
34. Robert F. Futrell, *The United States Air Force in Korea 1950-1953*, Revised Edition, (Washington, DC: Office of Air Force History United States Air Force, 1983), 138-140.
35. Jan Forsgren, Aeroflight, https://www.aeroflight.co.uk/user/fleet/north-korea-

although they did shoot down one from the 19th Group in July near Seoul.[36] On a June 29 B-29 strike on North Korean-occupied Kimpo Field, 19th Bombardment Group's B-29 gunners shot down one Yak-9 and damaged another.[37]

This first combat loss of a B–29 seized the attention of the bomber world, especially the gunners. On July 12th, Far East Air Force's Superforts were dispatched to roam the North Korean People's Army (NKPA) lines of communications, dropping their thirty-five 500-pound bombs individually on bridges, tunnel entrances, road junctions plus any observed troop concentrations, supply dumps, truck convoys, and even individual tanks. Three Yak–9Ps — probably scrambling from Suwon — intercepted one of these B–29s near Seoul. North Korean People's Air Force pilot Kim Gi-Ok (or less likely, Lee Dopn-Gyu — both claimed B–29 kills that day) shot out the number three engine with his cannon.[38] Ablaze, the B–29 escaped out to sea where the crew bailed out. Two of the crew were captured by the North Koreans. The remaining 11 men were rescued by the British frigate *HMS Alacrity*.[39]

One week later, three Yak–9Ps intercepted one of 10 B–29s dispatched to bomb bridges spanning the Han River. Catching up to a bomber near Seoul, the Yaks caused severe damage on the bomber

---

afyakovlev-yak-9-frank.htm
36. Robert F. Futrell, *The United States Air Force in Korea 1950-1953*, Revised Edition, (Washington, DC: Office of Air Force History United States Air Force, 1983), 99.
37. Douglas C. Dildy "The Korean People's Air Force in the Fatherland Liberations War: Part II," *Air Power History,* Vol. 59 No. 4: Winter 2012, 6. https://www.afhistory.org/airpowerhistory/Air_Power_History_2012_winter.pdf
38. Douglas C. Dildy "The Korean People's Air Force in the Fatherland Liberations War: Part II," *Air Power History,* Vol. 59 No. 4: Winter 2012, 9. https://www.afhistory.org/airpowerhistory/Air_Power_History_2012_winter.pdf
39. Douglas C. Dildy "The Korean People's Air Force in the Fatherland Liberations War: Part II," *Air Power History,* Vol. 59 No. 4: Winter 2012, 9. https://www.afhistory.org/airpowerhistory/Air_Power_History_2012_winter.pdf

and wounded the pilot.⁴⁰ The crew safely landed the bomber at its base at Kadena with over 100 bullet holes in it from the North Korean attack.⁴¹

The badly beaten North Korean air force made only a few sporadic, desultory appearances in August, most notably when a fighter (reported as a Lavochkin "La–5," a Soviet World War II-era fighter) attempted to attack a 307th Bombardment Group B-29 on August 15th. But the B-29's tail gunner drove off the Red fighter with two bursts from the bomber's tail guns.⁴² By August 10, 1950, the Korean Peoples Air Force no longer flew combat missions.⁴³

In September, the wartime increased flow of personnel into the 19th Bombardment Wing resulted in a significant overstrength for the organization. Far East Air Forces Headquarters ordered the transfer of 418 enlisted Airmen to other units that were understrength.⁴⁴ When the 307th Bombardment Group arrived at Kadena from MacDill Air Force Base, Florida, on August 1, Staff Sergeant Aaronson was transferred from the 19th Group to the 307th, which assigned him to be central fire control gunner for B-29 tail number 45-61814.

The U.S. Air Force hit every strategic target in North Korea. The North Korean Army was just about finished off as United Nations

---

40. Douglas C. Dildy "The Korean People's Air Force in the Fatherland Liberations War: Part II," *Air Power History*, Vol. 59 No. 4: Winter 2012, 9. https://www.afhistory.org/airpowerhistory/Air_Power_History_2012_winter.pdf
41. Robert F. Dorr, *B-29 Superfortress Units of the Korean War*, (Oxford, GB: Osprey Publishing, 2003), 14-15.
42. Douglas C. Dildy "The Korean People's Air Force in the Fatherland Liberations War: Part II," *Air Power History*, Vol. 59 No. 4: Winter 2012, 11. https://www.afhistory.org/airpowerhistory/Air_Power_History_2012_winter.pdf
43. Douglas C. Dildy "The Korean People's Air Force in the Fatherland Liberations War: Part II," *Air Power History*, Vol. 59 No. 4: Winter 2012, 4–13. https://www.afhistory.org/airpowerhistory/Air_Power_History_2012_winter.pdf
44. 19th Bombardment Wing Unit History, September 1950, AFHRA Reel M0505, 977.

forces closed in on the Yalu River in late October 1950.⁴⁵ With its mission nearly complete, the 307th stood down and began packing to return home from its almost 60-day temporary duty.⁴⁶

Suddenly, 30 divisions, out of as many as 463,000 Communist Chinese troops that had assembled for an offensive, infiltrated across the Yalu River and attacked in force.⁴⁷ Once again, U.S. and allied forces conducted a hasty withdrawal to the south. Bomber Command rescinded the standdown and combat crews prepared for a mission on November 1, 1950.⁴⁸

The 307th's target for this mission was the border town of Sinuiju, with its huge stockpiles of munitions and supplies, as well as the Chinese troops assembling to join the fight to the south. General MacArthur authorized the use of incendiary bombs to destroy these forces threatening, once again, to overrun the Korean Peninsula. And he directed Bomber Command to drive the crews to exhaustion if necessary.⁴⁹

The 307th was scheduled for another mission on November 9, but bad weather in the target area and a typhoon closing in on Okinawa forced a 24-hour delay. The next day, the aircraft were to fly from Kadena to the target then recover at Yokota Air Base near Tokyo

---

45. Robert F. Futrell, *The United States Air Force in Korea 1950-1953*, Revised Edition, (Washington, DC: Office of Air Force History United States Air Force, 1983), 205-207.
46. 307th Bombardment Wing History, October 1950, AFHRA Reel N064, p. 1135, 1146, 1155.
47. Eliot Cohen, "The Chinese Intervention in Korea 1950," CIA Reading Room November 1, 1988, C00616634 51. https://www.cia.gov/readingroom/docs/1988-11-01.pdf#:~:text=Thirty%20divisions%20of%20Chinese%20troops%2C%20rather%20than%20the,charts%2C%20attacked%20the%20UN%20forces%20with%20appalling%20effect.
48. 307th Bombardment Wing History, October 1950, AFHRA Reel N064, p. 1245.
49. Robert F. Futrell, *The United States Air Force in Korea 1950-1953*, Revised Edition, (Washington, DC: Office of Air Force History United States Air Force, 1983), 221.

to wait out the storm.⁵⁰ The target this time was an array of supply and troop concentrations hidden within houses and buildings in the ancient town of Uiju.⁵¹ Located on the Yalu River, Uiju had been a frequent battleground over the centuries of conflict in Korea involving Japan, China, and Russia. The mission would challenge the crews' airmanship, flying in a tight defensive formation.

As Philip Aaronson and his fellow crewmembers on 814 readied for the mission, a Protestant Chaplain offered the men a word of encouragement, that they were fighting for a righteous cause. He led them in prayer for a successful mission and safe return. On the flight line, a Jewish Chaplain met them on 814's hardstand and offered a traditional Jewish blessing.⁵²

The flight to the Initial Point, where the straight and level bomb run began, was uneventful. Once the formation turned to the southwest for the bomb run, pilots had to wrestle their huge Boeing Superfortresses against wicked crosswinds while carrying out high-altitude daylight bombing. The Yalu River presented a tortuous "no-fly" line as President Truman had declared U.S. aircraft would not fly beyond it into China under any circumstances for fear of triggering a third world war, now in the atomic age.⁵³

The Airmen of the 307th didn't realize they were about to tangle with more than 250 guns directly covering their bomb run on November 10,⁵⁴ and an entire Soviet Air Division of about 30 MiG-15s

---

50. 307th Bombardment Wing History, October 1950, AFHRA Reel N064, 1246.
51. 307th Bombardment Wing History, October 1950, AFHRA Reel N064, 1328.
52. Daniel B. Jorgensen, Chaplain, Major, USAF, *Air Force Chaplain's History Volume II 1947-1960*, (Washington DC: Office, Chief of Air Force Chaplains, 1961), 181. https://archive.org/details/airforcechaplain0002dani/page/n3/mode/2up
53. Robert F. Futrell, *The United States Air Force in Korea 1950-1953*, Revised Edition, (Washington, DC: Office of Air Force History United States Air Force, 1983), 220-223.
54. N.L. Volkovsky, Chief Editor, The War in Korea, 1950-1953, Chapter 12, "*Organization of Air Defense of Objects in the Rear and Ways to Inform Air Defense Artillery Assets,*" St Petersburg, RF: Military History Library Izdatel'stvo Poligon, 2000. http://

was now based at three Chinese bases nearby.⁵⁵ The 307th would soon become the first Bombardment Group to discover "MiG Alley."

As his aircraft hit the Initial Point, Philip diligently searched the skies for any sign of an airborne enemy, as Aircraft Commander 1st Lt Frank Johnston, radar operator Major David MacGhee, and bombardier 1st Lt Lyle Dodd, controlled the aircraft for the 10-minute straight and level bomb run. During this time, all other crewmembers were to remain strictly silent on the interphone so that the three men putting bombs on target could coordinate the intricate adjustments to keep 814 on track to put bombs on their aimpoint.

1st Lt Johnston and pilot 1st Lt Billy Foshee could not steer the big airplane along the designated bomb run, a northeast to southwest line down a wide avenue in the middle of Uiju. None of the 307th's crews in the nine-ship formation could handle the devilish crosswinds that blew them clear across the Yalu River. The lead ship, with Squadron Commander LtCol David G. Alford at the controls, decided to hold bombs, circle around, and make a second pass.⁵⁶ That was an unusual decision. It would give enemy gunners an extraordinary advantage at laying their guns on the bombers at their most vulnerable point, flying straight and level for five to ten minutes to get a good sight picture on the target. Thus far, though, the flak was minimal and MiGs were a rare sight.

When they reached the Initial Point for the second pass, enemy flak opened fire with vengeance. For the first time in his military career, Philip saw the black puffs of smoke with red bursts in the center as flak explosions came deadly close to his bubble atop the venera-

---

www.korean-war.com/Russia/KoreaPoligon541-573.html
55. Steven J. Zaloga, "Th Russians in MiG Alley," *Air and Space Forces Magazine* February 1, 1991. https://www.airandspaceforces.com/article/0291russians/
56. 6004th Air Intel Sv Sq, *Air Intelligence Information Report*, "Subject Interrogation of a Returned USAF Prisoner of War – Robert E. Burke, Capt., USAF," 8 Sep 53, 5. NARA RG 341 Series P 268-A Box 1 Loc 170-63-26-06 Folder: Burke, Robert E.

ble Superfortress. His three gunners and their 20-some calibre .50 machine guns would have zero effect on those thousands of artillery shells exploding in the air right next to their faces.[57]

Radar Operator Major David MacGhee jumped on the intercom: "Radar: Six bandits five miles out, coming in six o'clock low, 500 knots!"

Six MiG-15 jet fighters suddenly appeared out of nowhere.[58] Two attacked 814 from blind spots behind and below, pounding the B-29 with well-aimed 23mm and 37mm autocannon shells.[59] It all happened fast, but Philip, from his perch atop the fuselage in the Central Fire Control Bubble, witnessed the entire encounter as if it was in slow motion.

He was first to catch sight of the enemy. "Bandits two miles out! Coming up fast!" he barked into the intercom as he centered the lead MiG in his sight. The computer-controlled guns of the upper turrets swerved to the calculated intercept point. He could also see four other MiGs circling above like vultures waiting to feast on the expected kill. Side gunners S/Sgt William Brendle and Sgt Victor Foote both could see two MiGs beginning their sweep up from below and trained their computer controlled .50 calibre machine gun sights to track and lead the enemy fighters with the lower turrets.

But the MiGs were moving faster than anything they had ever trained on in gunnery exercises at their Guam or Florida home base.

---

57. BOMCOM COMBAT OPERATIONS CENTER DAILY DIARY 10 November 1950, Extract from Suppl Ltr Report fr 307th Bombardment Group (M), APO 239-1 /mca/ (n.d.) (Author Copy)
58. 6004th Air Intel Sv Sq, *Air Intelligence Information Report*, "Subject Interrogation of a Returned USAF Prisoner of War – Philipp (NMI) Aaronson, S/Sgt, USAF," Sep 53, p.6. NARA RG 341 Series P 268-A Box 1 Loc 170-63-26-06 Folder: Aaronson, Philip.
59. HQFEAF INCOMING MESSAGE FROM HQ USAF AFP-MP-12-E-3 TO: CG FEAF TOKYO JAPAN NR: AFPMPP-12-E (CASE #64) 13 Aug 51 (Author Copy)

Although the B-29's computer automatically could apply exactly the right amount of lead, the gunner still had to know the wingspan of the attacking fighter, dial it in, then rotate the aiming circle knobs and follow the target, keeping it centered so the computer could do the math. Even with the rapid calculations of the analog computer, the turrets were not fast enough to keep up with this jet flying twice the speed that the B-29 Remote Control Turrets were designed to track.

These nasty Russians were moving too fast, and no one had any idea of the wingspan of the swept-wing raptors coming their way. 814's four gunners tracked as best they could. Tail Gunner Corporal William Rose had, literally, a bird's eye view of the oncoming jets. He probably had the best chance of knocking them down with his powerful 20mm cannon, but the MiGs were so low by now he could not quite get his sight on them to track with a proper lead.

William and Philip punched their trigger buttons as they saw first one then the other MiG sweep up and out as they intersected 814's flight path. Philip had anticipated their maneuver and had his sight dialed in for 35 feet wingspan—a pretty good estimate, it would turn out. But it was to no avail. He watched helplessly as his rounds trailed the leading MiG. The entire crew heard a distinct "thunk, thunk, thunk," as the MiG's 37mm shells penetrated 814's skin somewhere aft of center.

It was over in less than a minute. The computer-controlled calibre fifties could not keep up with the nearly supersonic Soviet jets that flew more than twice the speed of the World War II Japanese propeller driven fighters that B-29 guns were designed to fight. Russian versions of this encounter claim the MiGs made a second pass.[60] No one of the repatriated crew mention it, perhaps because they were already going through the bail-out drill.

The enemy shells hit a fuel tank, the number two engine, tore up the

---

60. Krylov, 19-20.

left wing, and penetrated the radar compartment and the non-pressurized aft section.[61] Number two engine caught fire.[62] The rounds also destroyed aileron control of the ship. Without these hinged panels on the outside of the trailing edge of the wings, the giant plane would soon go into an uncontrollable roll, screwing itself into the ground. From his high perch, Philip could see that an engine was on fire and alerted 1st Lt Johnston over the intercom.[63]

Aircraft Commander 1st Lt Johnston was already aware that 814 would soon be lost, but he could still save his men. He rang the alarm bell three times and ordered the crew to prepare to bail. With pilot 1st Lt Billy Foshee also working the controls, and in perfect coordination with flight engineer Tech Sgt James Edwards, they initiated the controlled descent of the fatally injured 60-ton behemoth from 20,000 feet to the bail out altitude of 10,000 feet. Philip, in proper sequence as drilled, acknowledged with "Roger," over the intercom.

Philip disconnected and was quick to ratchet his seat down into the aft cabin that was already depressurized for the bomb run. He checked his parachute, his "Mae West" inflatable life vest, first aid kit, and survival packet. When the aircraft reached 10,000 feet, 1st Lt. Johnston rang the alarm one final time and ordered the crew to bail out. The bomb bay doors were the bailout location for the three gunners in the aft section. But the doors were not open, and the full bomb load was still hanging in the bomb racks. They would have to

---

61. 6004th Air Intel Sv Sq, *Air Intelligence Information Report*, "Subject Interrogation of a Returned USAF Prisoner of War – Robert E. Burke, Capt., USAF," 8 Sep 53, 2. NARA RG 341 Series P 268-A Box 1 Loc 170-63-26-06 Folder: Burke, Robert E.
62. 6004th Air Intel Sv Sq, *Air Intelligence Information Report*, "Subject Interrogation of a Returned USAF Prisoner of War – Philipp (NMI) Aaronson, S/Sgt, USAF," Sep 53, 2. NARA RG 341 Series P 268-A Box 1 Loc 170-63-26-06 Folder: Aaronson, Philip.
63. BOMCOM COMBAT OPERATIONS CENTER DAILY DIARY 10 November 1950, Extract from Suppl Ltr Report fr 307th Bombardment Group (M), APO 239-1 /mca/ (n.d.) (Author Copy)

make their way to the smaller aft hatch and squeeze out the narrow opening one-by-one[64].

Following bailout drill practices, the central fire control gunner would be the first man out. As he made his way to the hatch, Philip could see the snarled control cables that had been severed by the MiG-15's cannon rounds. Reaching the hatch, he turned the handle and let the door fly open in the wind. Exactly as trained, Philip faced the direction of flight, crouched, and somersaulted himself out the aft exit door.

He delayed opening his 'chute for 10 seconds rather than the standard four-count. He wanted to make sure he was clear of the doomed B-29. He was down to 7,000 feet when his canopy opened. He looked up to check the rigging. It was fully open and none of the lines were twisted. "Good to go," he said out loud with a big exhale, relieving the stress of the moment.

He could see 10 other parachutes above him. "They must have opened right away. Glad they made it," he thought. He looked below and saw a solitary human shape plummeting to the ground. He watched the body bounce one time, then lie deathly still. He would later learn from crewmates in POW camps that it must have been Staff Sergeant Augustine "Gus" Hendricks, the Electronic Countermeasures Operator.[65]

Philip looked up again and saw the B-29 flip to an inverted attitude, spiraling through the air, flames roaring, smoke billowing. Suddenly, the air quaked. Kaboom! A thunderous, mighty, reverberating roar as the engine fire must have reached fuel tanks and napalm bombs. The gigantic airplane instantly shattered into bits of metal and

---

64. Headquarters, Army Air Forces, Office of Flying Safety, *The B-29 Airplane Commander Training Manual for the Superfortress*, 1 February 1945, Reprinted 1 March 1945 with additions. 153.
65. 6004th Air Intel Sv Sq, *Air Intelligence Information Report*, "Subject Interrogation of a Returned USAF Prisoner of War – Robert E. Burke, Capt., USAF," 8 Sep 53, 2. NARA RG 341 Series P 268-A Box 1 Loc 170-63-26-06 Folder: Burke, Robert E.

debris.⁶⁶ In an instant, 814 was gone. The Russians credited Lieutenant Yu. I. Akimov with their first confirmed MiG-15 kill of a B-29 over North Korea.⁶⁷

Philip turned his attention to the ground now rushing up towards him. He was drifting just beyond a large body of water, a reservoir. He could see a town off to the east and a forest to the west. He pulled a two-riser "slip" to push air out one side of his open chute to steer his drift, somewhat, toward a small clearing in the wooded hills that now surrounded him. But he could not steer away from the human hazards waiting on the ground with guns already blazing up at him.

The moment before impact, he pulled himself up on the risers and flattened his forearms along his neck to protect his throat from the branches and trees surrounding him. He hit the ground with his feet and knees together, then instantly let his legs flex like a spring absorbing his impact, rolling on his feet to the side, and slamming the ground with the side of his calf, thigh, butt, hip and back, just as he had done in so many ground drills. He quickly stood up and checked his body. Nothing was injured. He felt lucky, blessed, really. As he gathered up the nylon ripstop-stitched canopy that had eased him safely to the ground, he saw another crewmember glide down to a landing about a quarter mile away. Philip stashed the chute among some brush and headed in the direction of his crewmate.

He was intercepted by North Korean soldiers who fired shots over his head to make sure he knew he was captured.⁶⁸ Enraged North

---

66. 6004th Air Intel Sv Sq, *Air Intelligence Information Report*, "Subject Interrogation of a Returned USAF Prisoner of War – Philipp (NMI) Aaronson, S/Sgt, USAF," Sep 53, 7. NARA RG 341 Series P 268-A Box 1 Loc 170-63-26-06 Folder: Aaronson, Philip.
67. Leonid Krylov & Yuri Tepsurkaev, *The Last War of the Superfortresses: MiG-15 vs B-29 Over Korea*, (West Midlands, England: Helion & Company, 2016), 19-20.
68. 6004th Air Intel Sv Sq, *Air Intelligence Information Report*, "Subject Interrogation of a Returned USAF Prisoner of War – Philipp (NMI) Aaronson, S/Sgt, USAF," Sep 53, 7. NARA RG 341 Series P 268-A Box 1 Loc 170-63-26-06

Korean civilians surrounded him, threatening with rocks and sharp farming implements. Staff Sergeant Philip Aaronson held his hands up in the air in the universal signal of surrender, formally becoming a prisoner of war. The North Korean soldiers quickly took his pistol and motioned for him to remove his parachute harness. They made him empty his pockets and took his wristwatch.

It was just after noon. The sun was out, but the air was cold. Really cold. It would turn out to be the coldest winter on record in North Korea. Philip had left his heavy jacket on board the B-29.[69] Since the cabin was pressurized at altitude, the temperature inside the B-29 was warm enough to wear T-shirts. On the bomb run, however, when the compartment was depressurized and the men wore oxygen masks, the cold of 25,000 feet altitude was no worse than what Philip had experienced in an unpressurized B-17. Besides, up in its combat position, the central fire control gunner's seat was a cramped space for a gunner outfitted with a flak vest, steel helmet, and parachute.

His captors marched Philip to a schoolhouse and sat him down in a nearby stack of cornstalks. Another crew member was already there, Major MacGhee, the radar operator whose position was seated on the floor of the aircraft in the same aft compartment as the gunners. The soldiers shoved them both inside the schoolhouse and guarded the entrance.[70]

At nightfall, their captors marched them into the dark to destinations unknown. Major MacGhee's hands were tied behind his back and pulled up to his shoulders. He could hardly stand, much less walk. Each time he fell, the angry guards shoved their guns in his face and jerked the ropes to force him back up, prodding him to continue the march.[71]

Folder: Aaronson, Philip.
69. Affidavit of Major David F. MacGhee, subscribed and sworn to by Major Frank M. Finn, JAGC, 13 April 1954, 3. http://www.kpows.com/amazingveterans.html
70. Affidavit of Major David F. MacGhee, subscribed and sworn to by Major Frank M. Finn, JAGC, 13 April 1954, 2. http://www.kpows.com/amazingveterans.html
71. Affidavit of Major David F. MacGhee, subscribed and sworn to by Major Frank M. Finn, JAGC, 13 April 1954, 2. http://www.kpows.com/amazingveterans.html

At some point Philip blew up at the guards for their treatment and yelled at the major, "Please, sir, don't fall again they're going to kill you!" Philip's outburst apparently prompted the North Korean officer in charge to unbind MacGhee's hands. They marched on into the night, and on, and on, and on, and on. About 3 a.m. they were placed into a Korean house filled with women and children. It had a heated floor; that was a welcome relief.[72]

They slept until just after noon on what was now November 11 when they were fed a sparse meal of sorghum-like kaoliang. This grain is much like rice but requires thorough preparation and lengthy cooking times. Otherwise, it goes through the digestive tract like birdseed. At about three that afternoon, MacGhee and Aaronson were joined by a third member of their crew, Navigator Captain Robert Burke.[73]

Captain Burke was badly beaten. His hands, like Major MacGhee's, were tightly bound with wire that obviously cut off the circulation in his hands. He told his crewmates what happened. "Lieutenant Johnston drilled us, me and Sergeant Jimmy Sanders, the Radio Operator, to bail out through the forward bomb bay doors while the men on the flight deck – aircraft commander, pilot, bombardier and flight engineer, used the forward hatch. That way we'd all get out quicker. But the bomb bay doors would not open. So, all six of us had to use the smaller forward hatch. I rolled out first while Jimmy set the 'destruct' switches on the sensitive equipment.

Philip knew Jimmy Sanders well. Noncommissioned officers generally were closer to each other than to the officers. Jimmy was an unusual airman. He had enlisted in World War II just a few days after Pearl Harbor in December 1941, but for some reason the Army Air Forces did not send him to Radio Operator School until late in

---

72. Affidavit of Major David F. MacGhee, subscribed and sworn to by Major Frank M. Finn, JAGC, 13 April 1954, 2-3. http://www.kpows.com/amazingveterans.html
73. Affidavit of Major David F. MacGhee, subscribed and sworn to by Major Frank M. Finn, JAGC, 13 April 1954, 3. http://www.kpows.com/amazingveterans.html

1944. In those days radio operators also had to qualify as gunners, and Jimmy Sanders was now one of the best radio operator/gunners in the Air Force. Just like Philip, he arrived at his combat unit in World War II days before the war ended. They were veterans, but neither had any combat missions before the Korean War.

Captain Burke continued.

On the way down, I counted 11 open canopies. I saw one crew member hit the after portion of the plane and most probably broke his neck. His canopy opened, but he just dangled there, not moving at all until he drifted down to earth. One man's 'chute must have failed. I watched him plummet all the way down 'till he bounced off the ground.

Philip wondered who that might be. Since it was someone who hit the rear of the aircraft it was probably someone who had bailed out of the same aft section that he had bailed out of himself. After the war, he learned it was most likely Side Gunner Staff Sergeant Dillman Brendle.[74] Like Philip, Dillman was from the 19th Bombardment Wing on Guam, now on temporary duty to the 307th. That was a huge tragedy: Dillman had a wife and two sons back in Tampa, Florida, just outside MacDill Air Force Base.[75]

Captain Burke said he landed on a mountaintop, slamming into the face of a cliff. "I was conscious but stunned. I turned my ankle and felt like I broke my back." He saw several North Korean soldiers making their way through the forest in his direction, so he headed south along the ridgeline, limping and slow. Several children spotted him but quickly moved on.

---

74. Defense POW/MIA Accounting Agency, Personnel Profile, SSGT Dillman Lawrence Brendle, Unaccounted For. https://dpaa.secure.force.com/dpaaProfile?id=a0Jt0000006lrL8EAI Brindle

75. "Boy, 8 Keeps Asking for His Daddy, But His Name Was Not on Prisoner List," *Tampa Tribune*, December 23, 1951, 1.

I tried to change direction and move down the slope, but the kids must have told the soldiers where I was, because next thing I knew the NK's showed up right in front of me, firing their rifles over my head. There was no escape. So, I surrendered. And here I am. Man, am I glad to see you two![76]

It was November 12, two days after they bailed out on their fateful flight. In these cold, wet woods of communist North Korea, they might as well have landed on another planet.

That afternoon, an English-speaking Chinese officer told them they would be moving by truck to another camp. Then he launched a long harangue on the glorious traditions of the Communist Chinese Volunteers of the People's Liberation Army. It was the American airmen's introduction to the enemy that had invaded across the Yalu River and forced the 307th to stay in the war just as they were set to go home in time for the holidays.

As darkness fell, the Chinese officer ordered Major MacGhee to give up his jacket to Philip. He helped the major, whose hands had not returned to full use after being bound so tightly for nearly two days. The North Korean soldiers marched them about 10 miles in a heavy snow across the frozen ground to meet up with a truck that would take them to another, larger POW camp. Along the way, the Chinese came up with one of their own ragged cotton-padded coats and made Philip return Major MacGhee his jacket[77].

The group halted their march at a spot near a town that Captain Burke recognized from aerial navigation charts as Taechon. The three

---

76. 6004th Air Intel Sv Sq, *Air Intelligence Information Report*, "Subject Interrogation of a Returned USAF Prisoner of War – Robert E. Burke, Capt., USAF," 8 Sep 53, 6-7. NARA RG 341 Series P 268-A Box 1 Loc 170-63-26-06 Folder: Burke, Robert E.
77. Affidavit of Major David F. MacGhee, subscribed and sworn to by Major Frank M. Finn, JAGC, 13 April 1954, 3. http://www.kpows.com/amazingveterans.html.

American airmen clambered up the tailgate of an old, rusted, Soviet cargo truck, joining about 20 rifle-toting North Korean soldiers already crammed into the cargo bed. The truck's engine coughed and sputtered while grinding its gears amid the smell of a burning clutch.

After a cold, dark, three-hour ride with the grimy, smelly North Korean soldiers, the three American airmen disembarked at a railroad siding just beyond a traffic circle with a tall radio antenna in the center.[78] It was 5 a.m. on November 13. Somehow the truck survived to arrive before dawn at a mining camp.

Major MacGhee placed this mining camp near the towns of Kunu-Ri and Huichon.[79] But Kunu-ri and Huichon are about 40 miles apart. Captain Burke figured they had travelled about 60 miles south of where they had been shot down near Kusong, a distance that would place them near Unsan.[80] As a navigator, Captain Burke most likely had the better perspective on the location of this "mining camp" POW holding facility.

Unsan was the site of one of the first battles between U.S. ground forces and the Chinese. Hundreds of American POWs were temporarily housed in this mining camp.[81] The Chinese Army during their early November assault across the Yalu River[82] set up a temporary

---

78. Affidavit of Major David F. MacGhee, subscribed and sworn to by Major Frank M. Finn, JAGC, 13 April 1954, 4. http://www.kpows.com/amazingveterans.html.
79. Affidavit of Major David F. MacGhee, subscribed and sworn to by Major Frank M. Finn, JAGC, 13 April 1954, 3. http://www.kpows.com/amazingveterans.html.
80. 6004th Air Intel Sv Sq, *Air Intelligence Information Report*, "Subject Interrogation of a Returned USAF Prisoner of War – Robert E. Burke, Capt., USAF," 8 Sep 53, 2. NARA RG 341 Series P 268-A Box 1 Loc 170-63-26-06 Folder: Burke, Robert E.
81. William Clark Latham Jr., *Cold Days in Hell: American POWs in Korea* (College Station, TX: Texas A&M University Press, 2012), 118-121.
82. Richard W. Stewart, *The Korean War: The Chinese Intervention, 3 November 1950 – 24 January 1951*, (Fort Leslie J. McNair, D.C.: U.S. Army Center of Military History, 2000) CMH Pub 19-8, 3. https://history.army.mil/html/books/019/19-8/CMH_Pub_19-8.pdf

POW compound at the mining camp after capturing hundreds of American soldiers, mostly from the 1st Cavalry Division's 8th Regiment, at Unsan.

In the pre-dawn darkness, the B-29 crewmen could make out a large, bare wood building. Philip noticed something odd about it. Not only was it not painted, but the slats ran up-and-down, not horizontally like a typical structure back in the States. Evidently it had been hastily and sloppily constructed. They also noticed several parked sedans with Soviet military markings.[83] The three Airmen were moved into a Korean house that was already crowded with eight U.S. Army soldiers who had been captured in the Unsan battle.

As the sun rose on the cold November morning, Philip could take in the entirety of his new North Korean prison camp. He stood at the entrance to a valley that stretched for several miles to the west. Though no fence surrounded the compound, it was walled-in by steep mountains that had been carved up by excavation. The Chinese were obviously in charge, ordering around the dozens of North Korean soldiers guarding and herding their American captives.[84]

That night, the Chinese took Major MacGhee away for what Philip figured was an interrogation. He wondered how he would do when it was his turn. Air Force Basic Training in World War II had instilled in Philip that he was required by the Geneva Convention to tell his captors only his name, rank, and service number. They also had an escape and evasion lecture by intelligence officers who came down to Okinawa from Tokyo.

Up to this time in the Korean War, only one B-29 had been shot down and two of the crew had been taken prisoner. Rumors were rampant on Kadena that there would be all kinds of Chinese torture

---

83. Affidavit of Major David F. MacGhee, subscribed and sworn to by Major Frank M. Finn, JAGC, 13 April 1954, 3-4. http://www.kpows.com/amazingveterans.html.
84. William Clark Latham Jr., *Cold Days in Hell: American POWs in Korea* (College Station, TX: Texas A&M University Press, 2012.) 118-121.

if they were captured. So far, while ordinary Korean citizens were hostile, Philip's captors had restrained them, and they did not subject him to anything he would call an interrogation. He resolved to resist to his utmost as he awaited learning of Major MacGhee's fate.

Major MacGhee returned to the prison home sometime after midnight. He huddled with Captain Burke and Philip to tell them what he had gone through. The story was hard to believe. The major said he was interviewed by the commanding general of all Chinese forces in North Korea, General Lin Piao.[85] During the session, Major MacGhee said, he had been kicked, slapped by the general, his personal guards, and the interpreter, and then threatened with execution as a war criminal. He refused to answer any military questions and was bound at the hands and taken away. As he was escorted out, the interpreter told him that unless he changed his attitude, he would never see his family again. On the way back to the hut, one of the guards fired a round at a large boulder right in front of Major MacGhee. Ten more communist soldiers hurried over and took him back to the headquarters building to the general's office.

According to MacGhee, General Lin Piao told him the shooting was a mistake and that he would be given an opportunity to learn the truth and would then join with the peace-loving peoples in the fight for permanent and lasting world peace.[86] They sent him back to rejoin

---

85. Affidavit of Major David F. MacGhee, subscribed and sworn to by Major Frank M. Finn, JAGC, 13 April 1954, 4. http://www.kpows.com/amazingveterans.html. (Author's note: This assertion is, at best, questionable. Declassified CIA reporting concludes General Lin was probably not present in Korea during the war. https://www.cia.gov/readingroom/docs/INTELLIGENCE%20REPORT%20LEADE%5B16026005%5D.pdf , 6.)

86. Affidavit of Major David F. MacGhee, subscribed and sworn to by Major Frank M. Finn, JAGC, 13 April 1954, 4. http://www.kpows.com/amazingveterans.html. (Author note: This episode with General Lin Piao does not appear in MacGhee's other interviews or interrogation records. He did not include it in his articles co-authored by Peter Kalischer for the January 22 and February 5, 1954 editions of Colliers magazine. While General Lin Piao may have been present with his troops,

Burke and Aaronson where, after relating the events, Major MacGhee told his crewmates that the threats to kill them as war criminals were all bluff.[87]

Camp life settled into a dismal routine for the next week. Then, on Thanksgiving Day, 1950, November 22, the North Koreans marched MacGhee, Burke, and Aaronson, eight other Americans, and 26 South Koreans out of the mining camp to a trail that took them generally southward. Starting out at about 4:30 in the afternoon, they marched around one of the mountains towering over the valley until they reached a crossroad around 6 p.m. A few yards beyond the intersection stood a bridge spanning a rushing river.[88]

They halted there for several hours until a truck arrived in the middle of the night. The 37 men jammed into a Soviet Army truck built to carry 16. They rode in a northerly direction, hour after hour, up one side, then down the other side of a mountain.[89] As dawn approached, the truck came to the bank of a large river, where the men off-loaded to cross the river on a narrow pontoon bridge as the truck was ferried across on a small barge. They re-boarded the truck and drove to a nearby town where the guards had a short but animated discussion with local officials. The truck took them back to the crossing point, then drove farther north to another town. North Korean Army Camp Commander Major Kim Dok Song met them and directed the escorts to take the POWs into nearby huts.[90] Welcome to Sombakol.

---

and the Chinese and Russians paid very close attention to American Airmen, I have found no specific reference to indicate he personally interviewed POWs.
87. Affidavit of Major David F. MacGhee, subscribed and sworn to by Major Frank M. Finn, JAGC, 13 April 1954, 4. http://www.kpows.com/amazingveterans.html.
88. Affidavit of Major David F. MacGhee, subscribed and sworn to by Major Frank M. Finn, JAGC, 13 April 1954, 5. http://www.kpows.com/amazingveterans.html.
89. Major David F. MacGhee, USAF, with Peter Kalischer, "Some of Us Didn't Crack," Collier's, January 22, 1954, 82.
90. Affidavit of Major David F. MacGhee, subscribed and sworn to by Major Frank M. Finn, JAGC, 13 April 1954, 6. http://www.kpows.com/amazingveterans.html.

A Chinese officer ordered his North Korean soldiers to separate Philip from the two officers. They led Major MacGhee and Captain Burke to an officers' hut. A North Korean guard motioned Philip to move forward with him, deeper into the compound.

As he walked up the rising valley floor, Philip saw dozens of huts overflowing with frozen, wounded, exhausted, gritty, filthy, ragged American soldiers. Their hollow eyes stared as they looked at the first American airman they had ever seen. Philip, now clothed in a Chinese jacket, and apparently still well-nourished and uninjured, moved smartly before them at the muzzle of the communist soldier's rifle.

Behind the rows of POW huts, Philip could see a line of primitive-looking mounds rising a few feet above the ground with traditionally clothed Korean women, most ashen-faced, and naked children, huddling near openings that led inside. Evidently, the Chinese had forced the residents out of their huts to make room for the prisoners. The civilian shelters had been excavated about five feet into the ground and covered with whatever debris and rubble the Koreans could scrounge to shore up a mud roof. A typical family—man and wife with three children—occupied a space perhaps eight-feet square with a kitchen and a sleeping area that doubled as a living space. Few had any furniture. Blankets were rare.[91]

As he walked on, the POW huts were fewer, and the dugouts were closer to the trail. Chinese camp leaders kept captive officers segregated from the enlisted men. They placed Philip with a couple dozen Army NCOs in a three-room, unheated shack. They kept him there for the next month. Prisoners subsisted on about four cups a day of cooked cracked corn.[92] Every other day their captors gave them a small cup of soup made of greens and chunks of pig gut.

---

91. Taewoo Kim, "Overturned Time and Space: Drastic Changes in the Daily Lives of North Koreans During the Korean War," *Asian Journal of Peacebuilding* Vol 2. No. 2 (2014), 99.
92. William Clark Latham Jr., *Cold Days in Hell: American POWs in Korea* (College

When Philip got his cup of soup, he wolfed down the first warm liquid with meat he'd had in weeks. "What was that tasty, spongy looking meat?" he asked a fellow American. "Pig stomach," drawled a Southern sergeant. "It's supposedly a Chinese delicacy but it tastes just like my Gramma's chitlins!" Philip figured, under the circumstances, God would excuse him from kosher rules in this time of desperation.

The camp had few latrines outdoors, none inside. When the three airmen arrived, all latrines were already overwhelmed with waste. The POWs had to get permission to exit the hut and do their business in the nearby wood. Their drinking water came down from the mountains into the stream running through the valley. By the time it reached Sombakol, the excesses from POW latrines and the ageless routines of the locals contaminated it with human waste. Diarrhea and dysentery soon set in.[93]

As time wore on and the meagre diet persisted, prisoners had a harder time getting outside before making a mess on themselves. Often the North Korean guards refused to let a man out for no reason other than contempt. The inevitable result was a smelly mess permeating their bodies and the entire hut—not to mention devilish lice colonies and other painful inflammations developing in the most sensitive areas of the skin. They had no provision for washing clothes in the perpetually frozen air. During his stay at Sombakol, Philip, too, became a smelly mess.

One measure of redemption at Sombakol was the 8th Cavalry's chaplain, Father (Captain) Emil Kapaun, who was imprisoned there. When the Chinese surrounded the 8th Cavalry in the Battle of Unsan, Father Kapaun was among the few survivors. He would often slip out of the officer's hut in the still of the cold night, steal nutritious food—vegetables, potatoes, fish, pork—from the unsuspecting

---

Station, TX: Texas A&M University Press, 2012), 119.
93. William Clark Latham, Jr., Cold Days in Hell: American POWs in Korea (College Station, TX: Texas A&M University Press, 2012) 119.

North Korean soldiers' storehouses, and bring it to the American enlisted men. He was never caught in the act at Sombakol.[94] In a few weeks, Philip would once again be blessed by the future Medal of Honor recipient.

While at Sombakol, Philip was never interrogated. That good fortune would soon change. He along with Major MacGhee and Captain Burke were among eight Americans that North Korean soldiers marched out of Sombakol on December 22. They trudged south through the cold snow for two days, covering about 50 miles. On Christmas Eve they arrived at a village near Pukchin. This was an interrogation center run by the Chinese military.

The POWs huddled in their hut that night and sang carols for Christmas 1950. The Aaronson family typically partook in their country's holidays, and Philip joined in solidarity with Christians MacGhee and Burke. Their New Year's celebration included a Christian prayer, followed by Philip's recital of a traditional Hebrew appeal for deliverance and blessing that he then translated for his fellow "Long Rangers", as the men of the 307th were known from their legendary long distance missions in World War II: "Deliver me, O, Lord, from the evil man; ... Thou art my God: hear the voice of my supplications, O Lord."[95]

Philip, the artiste of the group and deeply moved by the moment, asked his fellow POWs if anyone objected to him reciting "The Bluebird of Happiness." It was a popular song made famous by tenor Jan Peerce in a 1945 recording. MacGhee and Burke told him to go ahead. Philip poured out all the pent-up emotion these three Airmen had felt since their fateful November 10 mission:

---

94. William Clark Latham Jr., *Cold Days in Hell: American POWs in Korea*, (College Station, TX: Texas A&M University Press, 2012), 119.
95. Major David F. MacGhee, USAF, with Peter Kalischer, "In Korea's Hell Camps: Some of Us Didn't Crack," *Collier's* January 22, 1954, 84.

So be like I, hold your head up high 'til you find the bluebird of happiness.
You will find greater peace of mind, knowing there is a bluebird of happiness.
And when he sings to you, though you're deep in blue
You will see a ray of light creep through
And so remember this, life is no abyss
Somewhere there's a bluebird of happiness.[96]

Beginning in January, each POW was taken, one at a time, to an interrogation session. Chinese Lt Col "Dirty Pictures" Wong, a swaggering, arrogant, brutal, sadistic man, conducted the questioning. He was university educated, spoke good English, and had served with American troops in China in World War II.[97] He earned the nickname "Dirty Pictures" because he collected lewd photographs and displayed them around his interrogation rooms for his own personal pleasures. Some he had confiscated from captive Americans.

North Korean soldiers bound Philip's hands behind his back and hustled him into the headquarters building after dark. Wong stuck his face inches from Philip's nose. As soon as the Chinese lieutenant colonel opened his mouth, the stench of raw fish and garlic blew the stoic expression off the face of the American from downtown Harrisburg.

"What Group are you in?" demanded Wong. "Answer me!"

Philip remained silent.

Wong reared his arm back and smacked Philip with a left-handed power slap.

Through the blurred vision and ringing in his ears, Philip blurted,

---

96. Musixmatch, "Lyrics of Bluebird of Happiness, Jan Peerce," https://www.musixmatch.com/lyrics/Jan-Peerce/Bluebird-of-Happiness

97. Army Security Center 8589th AAU Fort George G. Meade, Maryland, Interview Report, Mac Ghee, David F. ASCIR #0067 28 June 1954, 35.

"Philip Aaronson, Staff Sergeant United States Air Force, AF FIVE EIGHT ONE EIGHT EIGHT TWO!"

Wong's brown skin turned blaze orange as he turned his shoulders to his right and hauled off with his strong hand, this time clenched in a fist. The blow knocked Aaronson to the floor. Dazed, but conscious, Philip got up, stood tall, stared ahead, and said nothing.

The interrogator unholstered his pistol, slowly raised it to Aaronson's eye level, and aimed the barrel at the wall directly to the rear.

"Young man, your life depends on the answers you give me. We are not signatories to the Geneva Convention. You'd better answer.[98] What is your Group?"

Silence.

"Now, Aaronson AF five eight one eight eight two, unless you talk, I will kill you."

Phillip braced himself. By this time, word had spread about Dirty Picture Wong's opening gambit. He knew the half-crazy officer was about to fire the weapon, but his practice was to shoot at the wall or ceiling, though close enough to shatter an eardrum. Without turning his head, Philip forced his eyeballs in a hard bank to starboard.

He could not see the gun's action, but he heard the distinct click of the hammer coming back to the cocked position. Wong was armed with the then-standard Chinese Type-51 semi-automatic pistol. To the POW gunner's eye, it looked a lot like an American M1911.[99] It was another classic John Browning design, just as were the B-29's machine guns.

Wong did not need to pull the hammer back to fire it off. But he did so for dramatic effect. Time slowed. In the microsecond it took for

---

98. Saul Kohler, "Return From Hell…" *The Harrisburg Patriot-News*, November 8, 1953. 81.
99. "Chinese Tokarev Pistols – Military and Commercial Models," *The American Rifleman*, posted on February 3, 2017. https://www.americanrifleman.org/articles/2017/2/3/chinese-tokarev-pistols-military-and-commercial-models/

Wong to squeeze the trigger, Philip slammed his eyelids shut so tight that his tiny tensor tympanic muscle set off a low-pitched rumble before the concussion even went off.

BANG! The soft thunder instantly turned into a mighty blast. It lasted only an instant, but the ringing and pain that followed were overwhelming.

Wong calmed down, as if shooting his gun released something inside him. He left the room, leaving Aaronson and the guard alone for about a half hour. When he returned, he was accompanied by the Camp Commander, General Wang. In an obvious bad cop–good cop routine, the General addressed the prisoner in a calm demeanor: "A -Ron -Son." His English was awkward and broken, "We apologize for lost temper. Peace-loving people do not do this things. You will learn truth here."

With that, the session was over. The guard took Philip back to his hut. The next day, Dirty Pictures summoned him again to his lewd interrogation room.

"Sergeant Aaronson, we know you are in the 307th Bombardment Wing from Kadena base. Draw us a picture of the runways and hangars."

Philip remained defiant. This time he tried to bluff his way through this soft-touch interview. He sat at the small table where his interrogators had placed a single sheet of paper and a fat No. 2 pencil. He sketched a roughly scale drawing of the 9,000-foot North-South runway at the fighter station on Guam, Naha Air Base. He knew it well from frequent visits while he was in the 19th Bombardment Wing stationed at nearby Anderson Air Base. Philip lay down his pencil and put his head down as if exhausted from the effort.

Dirty Pictures Wong picked up the drawing. He stared at it for a few seconds, then tore it up. He said nothing as he walked over to a shelf and pulled out a rolled up architectural drawing. He unrolled it on the desk before Aaronson, and Philip could see it was an official blueprint of the Kadena Air Base prepared by the U.S. Army Corps of Engineers. It was dated July 1950.

"Aaronson, next time you will tell us the truth. There are many large rocks here. If you lie to us one more time, we will put you under one and nobody will ever find out which one you are under."[100]

With that, the guards took Philip outside and stood him before a squad of soldiers going through their manual of arms. When the American gunner appeared, they went to the "order arms" position, resting their rifles' butts on the ground while standing at attention and grasping the weapons' forearms. Philip could not understand the Korean language, but the soldiers were unmistakably formed as a firing squad. "Ready"; the squad brought their rifles to port arms position. "Aim"; they brought their rifles up to the standing fire position.

Then Wong ordered, "Aaronson, RUN!"

Philip thought he had had it. He ran toward his hut. The soldiers did not shoot. "If they had shot me, they would have done me a favor," he recalled a few months after he was repatriated in 1953.[101] He resolved to attempt an escape that night.

He didn't have a thoroughly developed plan. He would find a moment in the wee hours of night when the guards were less attentive, make a dash for the forested hillside, and then figure out how to make his way south to friendly lines. He counted on that being only a couple nights' journey on foot.

It was about 3 a.m. when he eased himself up to the entrance of his hut. He managed to step over the other men without disturbing them. If anyone asked, he'd just tell them he needed to go out to the

---

100. Author's re-creation based on 6004th Air Intel Sv Sq, *Air Intelligence Information Report*, "Subject Interrogation of a Returned USAF Prisoner of War – Philipp (NMI) Aaronson, S/Sgt, USAF," Sep 53, 8, 21. NARA RG 341 Series P 268-A Box 1 of Loc 170-63-26-06 Folder: Aaronson, Philip.
101. Saul Kohler, "Return From Hell…" *The Harrisburg Patriot-News*, November 8, 1953. 81.

latrine. He peered left, right, and forward. All was clear. He leaped forward as he had done back at William Penn High School on the Junior Varsity football team in making a fullback rush.

Two steps later he was flat on his back. His congested lungs had collapsed from the exertion, and his atrophied leg muscles refused to propel him any farther. His attempt did not escape the notice of the guards, who sounded the alarm and quickly took him to an isolation chamber, a 4-foot-by-4-foot hole in the ground about six feet deep. They left him there for the remainder of the night and all the next day.[102]

Over the next week the Chinese interrogated him three more times. Each time they repeated the firing squad tactic. They asked him about the "AN6," something about which he had no clue. He made up a description of an obsolete radio that they were supposedly forced to use while the "modern" radios were taken out of the B-29s and installed in jet fighters. The Chinese called his bluff, pulling out an official Technical Order showing diagrams and specifications for the AN/APQ-23 High Altitude Bombing Radar.

Then they questioned him about Air Force Air Rescue procedures. On this, Philip was thoroughly informed from his training and drills both at Anderson and Kadena. He pled ignorance, claiming they were so secret that enlisted men were not allowed to know such things.

His final session was about Air Force pay scales. They wanted to know how much each enlisted grade made. On this, Philip figured they probably had some idea based on newspaper reporting. He drew out a table by grade but listed monthly pay at about a tenth of the actual amounts, figuring Chinese and North Korean pay was so low they would not believe how much American Airmen really made.

---

102. Saul Kohler, "Return From Hell…" *The Harrisburg Patriot-News*, November 8, 1953. 81.

"You lie, American Sergeant. No one makes that much money. What can you do with that much?"

"Well, many of my buddies own a car."

The communist was incredulous, but did not pursue the matter further.

That was Philip's last interrogation at Pukchin. On January 18, 1951, he, Burke, and MacGhee were moved to another Chinese interrogation center about 12 miles away.

The next morning, Philip was subjected to an interrogation that would soon be known around the globe. They eased him into a comfortable, heated room and politely seated him at a small table. His examiner was dressed in plain civilian clothes and introduced his note-taker, a European correspondent for a communist newspaper. Both spoke good English and began asking questions about his personal life in the United States.

"Philip," his questioner said. This sudden personal demeanor lowered Philip's guard. He was cold, tired, and hungry. Some of his teeth were broken from the cracked corn diet. His sore gums had started to swell. Maybe there was something to this notion of a Chinese "lenient" policy. "What was your life like in Harrisburg, before you joined the U.S. Army Air Force in 1943?"

How did they know he had served in World War II, before the U.S. Air Force had been separated from the Army? And how did they know he was from Harrisburg? But he nibbled at the bait anyway. "Well, I went to school. I played football. I acted in dramas." He stopped there, realizing he had already ventured beyond name, rank, and serial number.

The reporter then asked the questions: "What was your family like?"

"I have two brothers and five sisters. My father worked for grocery stores and my mother was a fine wife and mother. We always had plenty of food and were quite happy."

"Do you miss them? Do you think they know you are safe with us? Would you like to send them a letter?"

He did miss his family, sorely, and he wished he could go home. A letter would at least let them know he was alive.[103] "Yes, I'll write a letter home," he said.

This was the first time any of the prisoners had a chance to send a letter home. They brought out a sheet of plain paper, a pencil, and an envelope. He wrote a brief note to his father and mother assuring them he was safe, alive, and a prisoner of the Chinese, giving his love to all and asking them to tell a friend he was alive.

The newspaper man offered to mail it for him.

Philip folded the paper, slipped it into the envelope on which he wrote his home address:

Mr. And Mrs. Hyman Aaronson and Family
2128 North 3rd Street
Harrisburg, Pennsylvania

The correspondent took the envelope and assured Philip it would promptly be mailed to the United States.

Philip thought the letter would let Americans know that at least one of the prisoners of war was alive. It seemed to Philip that no harm would come of this to the military effort. And it would be good for a letter to find its way back to the States so that military authorities would have an indication that some POWs were still alive. He would come to regret going even a small step beyond name, rank, and serial number.

The next day, the Chinese had their Korean vassals move Aaronson, Burke, MacGhee, and about a half dozen other POWs to yet

---

103. "Main Interest of Freed City POW Is Just 'Being Home,'" *The Patriot-News*, Harrisburg, Pennsylvania, September 28, 1953, 20.

another camp a few hours away by foot. It was a larger camp, housing about a thousand Americans. That is, those who were still alive. As the three airmen of B-29 tail number 814 marched in toward their assigned hut, they passed uncounted numbers of dead Americans, heaped in "…stacks of bodies four or five high, with frozen arms and legs sticking out everywhere."[104] This camp at Pukchin Tarragon was known as Death Valley.

Death Valley was about 25 miles North of Pyongyang. It was one of three temporary POW collection points run by the North Koreans to accommodate the thousands of UN soldiers they had collected in November, mostly from the U.S. troops of the 2nd Infantry Division that had been surrounded at the Chongchon River Valley battle disasters.[105] The prisoners were housed in long, narrow, windowless, unheated, lice-infested buildings with a kitchen at one end. The camp got its name because POWs were dying by the dozens a day. Men gathered to pray, but English-speaking Chinese guards told them, "God can't help you now. Only Chinese volunteers can help you."[106]

As he walked into the camp, Philip had to cover his nose. A raw stink filled the air. The smell at Death Valley was putrid. And it never went away. It was from the unbelievably unsanitary conditions of the camp. The latrines, built decades ago when the site was a mining camp, had filled weeks earlier, soon after the camp housed its first occupants. The Chinese overseers, reluctantly, had the prisoners dig another one, smack in the middle of the mud-hut villages that housed a thousand

---

104. William Shadish, M.D. with Lewis Carlson, *When Hell Froze Over: The Memoir of a Korean War Combat Physician Who Spent 1010 Days in a Communist Prison Camp*, (New York, NY: iUniverse, Inc.,2007), 39.
105. Affidavit of Major David F. MacGhee, subscribed and sworn to by Major Frank M. Finn, JAGC, 13 April 1954. http://www.kpows.com/amazingveterans.html
106. Raymond B. Lech, *Broken Soldiers*, (Urbana, IL: University of Illinois Press, 2000), 42.

GIs. It, too, quickly overflowed its capacity, draining downslope across the area and into a stream.[107]

It was a rough night. Philip was placed in a room with 30 other enlisted men who had been there for weeks. These infantrymen from the 2nd Division[108] had developed a sleeping routine to cope with the overcrowding. The only geometry that allowed everyone to lay down was for all to sleep on their side. If for some reason you wanted to shift, you had to wake the man next to you and make him shift in the same direction. Then he had to wake the man next to him. And on it would go across the entire room.[109] The only other option was to remain all night on whatever side you fell off to sleep.

As Philip laid himself down on a ragged straw mat, hundreds of lice jumped onto his body and began burrowing their way into the few warm, moist places available on his frozen skin.[110]

The next morning, Philip was awakened by the movement of 26 of his 29 roommates to get up for chow: another four cups of cracked corn. Three men did not get up. Their corpses were now in an eternal sleep. The men next to them carried the bodies out onto the pile to freeze within an hour in the sub-zero air.

Phillip was famished after two days of marching. He bit down hard. Something other than a corn kernel was in his mouth: he spit out a tooth. Somehow, he still managed to finish the so-called meal.

Later that morning Philip was escorted to a building for interro-

---

107. William Shadish, M.D. with Lewis Carlson, *When Hell Froze Over: The Memoir of a Korean War Combat Physician Who spent 1010 Days in a Communist Prison Camp*, (New York, NY: iUniverse, Inc.,2007), 31.
108. William Latham, *Cold Days in Hell: American POWs in Korea*, (College Station, TX: Texas A&M University Press, 2012), 121.
109. Raymond B. Lech, *Broken Soldiers*, (Urbana, IL: University of Illinois Press, 2000), 40.
110. William Shadish, M.D. with Lewis Carlson, *When Hell Froze Over: The Memoir of a Korean War Combat Physician Who Spent 1010 Days in a Communist Prison Camp*, (New York, NY: iUniverse, Inc.,2007), 30.

gation. He noticed that, again, reporters were sitting in the interview room ready to take notes and photographs. To Philip, some of the reporters looked to be distinctly Russian. He resolved not to fall for their "good cop" routine again.

A Chinese lieutenant colonel was his interrogator, H.T. Hsiang. Americans called him "Shang." He spoke in a low voice. His English was perfect. "Sergeant Aaronson, you must answer our questions if we are to mail your letter to your family. Tell us about your B-29. What was its tail number?"

Philip was troubled. He knew he should resist this line of military questions. But he also believed that no other prisoner had been allowed to send home a letter as proof of life. He tried to evade, "I am just a gunner. They do not tell us such information."

"Aaronson, your plane was completely destroyed. It is unrecognizable. We need to know its serial number so we can properly inform authorities who can confirm your crew's survival."

"Shang" was clever. Philip wondered how his English got so good.

"In case you are wondering," the Chinese officer said, "I was with your OSS (Office of Strategic Services) in China during World War II while you were in England with the 96th Bomb Group."[111]

"Holy Cow, they must have a complete file on me," Philip thought to himself. "Why are they so interested in 814?" He decided he better not start down that trail. He did see the aircraft blow up into a million pieces. Maybe he should tell them so they could do what they said they would do. "No! these commies are never going to do the right thing," he concluded in his mind.

"If you know about me, you must know about my aircraft. They never told me what our tail number was," Philip replied.

"Sergeant Aaronson, you can trust me. Your General Donovan

---

111. Army Security Center 8589th AAU Fort George G. Meade, Maryland, Interview Report, Mac Ghee, David F. ASCIR #0067 28 June 1954, 32.

parachuted me behind Japanese lines to recruit Chinese people to help American airmen who were shot down.[112] I will help you while you are our guest. But you must help me."

Philip was tempted by Shang's professional manner. His question seemed genuine. But he had been seduced once before. They did not mail his letter as they said they would and, once again, here were reporters present for the interrogation. It all seemed to Philip to be another slippery slope of deception.

Suddenly, outside the building, rifle fire started popping off across the camp. Soldiers were hollering and screaming in high-pitched Chinese and Korean as the guards hurried to duck under whatever cover was nearby.

The room was empty. He was alone. He rushed out the door to see what was happening. The reason for the alarm then appeared in the sky. A B-26 bomber made a low pass over the camp. Fortunately, the pilot recognized that his intended target of opportunity was not what his mission briefing told him it would be, and he held fire as he zoomed by in what would have been a deadly strafing run.[113]

In a flash, Philip saw another opportunity for an escape attempt.[114] He ran through the doorway of the interrogation building. In front of him was the steepest mountain he had ever seen. He spun around. The mountain on the other side of the valley was even steeper. He glanced up the valley to the north. It looked like it stretched for a couple of

---

112. Author's reconstruction based on declassified intelligence on Lt Col H.T. Hsiang. Army Security Center 8589th AAU Fort George G. Meade, Maryland, Interview Report, Mac Ghee, David F. ASCIR #0067 28 June 1954, 32.
113. Author's re-creation based on 6004th Air Intel Sv Sq, *Air Intelligence Information Report*, "Subject Interrogation of a Returned USAF Prisoner of War – Philipp (NMI) Aaronson, S/Sgt, USAF," Sep 53, 7. NARA RG 341 Series P 268-A Box 1 of Loc 170-63-26-06 Folder: Aaronson, Philip. Major David F. MacGhee, USAF, with Peter Kalischer, "In Korea's Hell Camps: Some of Us Didn't Crack," *Collier's* January 22, 1954. 82.
114. Saul Kohler, "Return From Hell..." *The Harrisburg Patriot-News*, November 8, 1953, 81.

miles. But about a thousand yards away there was another POW encampment with guards swarming all over the place. To the south, at the entrance to the camp, guards were still under cover, fearful of another pass by the B-26.

He quickly chose the first mountain. It had the most timber and vegetation he could use to cover his route. He stepped out on the first move of his dash to freedom. In two steps he again collapsed to the ground. He could not get enough breath from his atrophied lungs to go another step. And his now pencil-thin rubbery legs gave way.

The Chinese commanders shouted orders, and the North Korean guards hustled the prisoners back into their miserable huts. For the next few days, the buzz among POWs was that a rescue mission was in the works. Surely that B-26 crew noticed the presence of so many Americans on his low pass. In small groups, they clandestinely made plans to take control of the camp from the communists once the mission arrived. This expectation gave them hope and a purpose that kept morale up.[115]

After just one more night in Death Valley, the Chinese formed up every man in the Camp. Nearly a thousand were going to march to yet another prison camp.[116] The camp commander told the assembled group they were going to a just-finished permanent camp along the Yalu River. He broke the prisoners into two groups. "Sergeant Aaronson," he ordered, "you will be in charge of the enlisted men."[117]

Philip had no idea why they selected him—and he was not impressed by the command selection. How was he going to keep these starving, injured, despised, and dejected men on this march into the

---

115. William Latham, *Cold Days in Hell: American POWs in Korea*, (College Station, TX: Texas A&M University Press, 2012) ,130.
116. William Latham, *Cold Days in Hell: American POWs in Korea*, (College Station, TX: Texas A&M University Press, 2012), 121.
117. Affidavit of Major David F. MacGhee, subscribed and sworn to by Major Frank M. Finn, JAGC, 13 April 1954, 10. http://www.kpows.com/amazingveterans.html

deadly cold night? No matter the route, the only direction of travel was up and down uncounted steep mountains.

At dusk on January 22nd, the formation moved out. "So long Death Valley," mused Philip to himself as he mouthed a passage from Tehillim: "May God shine a light unto our path." They left behind about 300 POWs who were too weak to travel. The communists left two captured American Army doctors to care for them. About six weeks later, only 109 survived to be transported out. Historian William Latham found that, as the POWs departed, using oxcarts to transport those who could not walk, "Starving dogs crept down from the hills to feed on the naked, unburied bodies of the dead American soldiers."[118]

For the crewmembers of B-29 tail number 814 and their fellow POWs, theirs was a Death March out of Death Valley. This move was different from previous POW camp transfers, however. In many instances, guards assisted POWs struggling to keep up. Perhaps the North Koreans were now suffering as much cognitively under their superior-minded Chinese allies as American POWs suffered physically under the brutal North Koreans.

Guards helped an American who was way behind on a steep upslope. One guard tied a rope around a U.S. soldier's mid-section and lashed the other end around his own waist. On each step, a gentle tug helped the struggling GI stay on his feet. Other communist soldiers followed closely behind the lame, wrapping their arms around an American to support them as they lurched and stumbled uphill.[119]

At night the formation stopped at a North Korean village where the Chinese, once again, forced local Koreans to vacate their homes

---

118. William Latham, *Cold Days in Hell: American POWs in Korea*, (College Station, TX: Texas A&M University Press, 2012), 124.
119. Affidavit of Major David F. MacGhee, subscribed and sworn to by Major Frank M. Finn, JAGC, 13 April 1954, 10. http://www.kpows.com/amazingveterans.html

to accommodate the POWs. Americans were jammed two dozen to a small room. Meals consisted of the new normal few handfuls of birdseed. Digestive tracts continued to spew forth their fetid waste uncontrollably.

Philip's formation covered about a hundred miles in seven days. The final 10 miles seemed to take forever, although this last segment was less circuitous and had fewer slopes to traverse than the previous week. Philip led from the front, but from time to time he made his way to various sections of the formation to get a feel for the condition of the men. He was exhausted and starving, but he tolerated the physical strain better than most of the others.

Along the way he got to know Tibor Rubin, a fellow Jew nicknamed "Teddy" from childhood. He was an extraordinary American soldier. He survived a Nazi concentration camp, resettled in the United States, and enlisted in the Army in 1950. He was assigned to the 1st Cavalry Division in Japan when war broke out. Rubin was captured at Unsan and was now on this journey to Pyoktong.[120] Philip found inspiration in the spirit and energy this proud American soldier openly displayed.

They arrived at Pyoktong on January 27. The camp was situated on a mile-long stretch of land surrounded on three sides by the Yalu River in a saddle formed by hills along the banks of the two long sides of the peninsula.[121] The camp itself was a hamlet of homes and shops that the Chinese seized to form a headquarters complex and POW compound. The Chinese also appropriated the village Buddhist Pagoda and the Presbyterian Church that had been established by indig-

---

120. Daniel M. Cohen, *Single-Handed: The Inspiring True Story of Tibor "Teddy" Rubin – Holocaust Survivor, Korean War Hero, and Medal of Honor Recipient*, (New York, NY: Berkley Caliber, 2015), 1-190.
121. Affidavit of Major David F. MacGhee, subscribed and sworn to by Major Frank M. Finn, JAGC, 13 April 1954. http://www.kpows.com/amazingveterans.html

enous Christians in the early 20th century.[122] Generations of Korean families in this town were forced out to make room for the Chinese occupiers and the incoming POWs. The Koreans moved to makeshift shelters, holes they dug in the ground, in nearby Pyoktong village.

Camp 5, Pyoktong[123]

Tibor Rubin noted that a row of wood and metal skeletons of buildings burned as they plodded through Main Street Pyoktong.[124] Angry peasants lined up to jeer as the men passed. Rocks flew. Their rage was directed at all American prisoners, though they had no way of knowing that it was Aaronson's 307th Bombardment Group that bombed Pyoktong with incendiaries on November 19, 1950.[125]

The prisoners were crowded into vacated villagers' huts, 30 men to an 8-by-10-foot room. The next morning, January 28, Philip was photographed and fingerprinted. At this camp, the Chinese military was clearly in charge. Most of the guards were so-called "Communist Chinese Volunteers" of the People's Liberation Army and North Korean soldiers were subordinate to their Chinese communist masters.

In the United States that day, the major wire agencies ran a remarkable story in newspapers across the country. Chinese Radio announcers read two letters they attributed to SSgt Aaronson over the air.[126] One was "an open letter to the combat crews of B-29 bombers that are operating over Korea and in general to all personnel of the Air

---

122. William H. Funchess, *Korea P.O.W.: A Thousand Days of Torment*, (Clemson, SC: South Carolina Military Museum, n.d.), 63.
123. Army Security Center 8589th AAU Fort George G. Meade, Maryland, Interview Report, Mac Ghee, David F. ASCIR #0067 28 June 1954.
124. Daniel M. Cohen, *Single-Handed: The Inspiring True Story of Tibor "Teddy" Rubin – Holocaust Survivor, Korean War Hero, and Medal of Honor Recipient*, (New York: Berkley Caliber, 2015), p. 210.
125. 307th Bombardment Wing History, October 1950, AFHRA Reel N064, p. 1328.
126. "Reds Use Alleged GI Letters As Propaganda," *The Baltimore Sun*, January 29, 1951.

Force" addressed to his fellow airmen.[127] The other letter was allegedly addressed to his parents in Harrisburg and urged Americans to give up the fight so that they all can come home.[128]

Hyman and Dorothy Aaronson had not seen any letters from Philip since early November. That letter was from his Kadena address. In that letter, he told them not much more than that he was a gunner on a B-29. The latest word they had received since was a telegram from the Defense Department reporting him as Missing in Action.

They prayed he was still alive, but this letter that Chinese Radio was broadcasting seemed like a cruel hoax. Hyman told the Associated Press that he did not believe it was real, "I don't believe Phil would write that kind of letter unless he was forced to do it…(he)…probably wrote a letter to let us know he was alive, and they added the propaganda."[129] He was correct.

In fact, they would not receive a letter from their POW son until August 1951. That later letter was dated February 11 and was in Philip's handwriting but obviously written under duress. He said the Chinese gave him enough to eat and that he could receive packages. He asked them to send hard candy, tobacco, and concentrated foods. He told them not to worry and that he would be released when the war was over.[130]

Camp 5 was Hell frozen over. The communists did not worry about escapes because there was nowhere the Americans could go, not to mention they were emaciated and incapable of walking very far anyway. The North Koreans provided wood for fires each day. But

---

127. "Hear Message from U.S. POW," Marion, IN *Chronicle Tribune*, January 29, 1951, 3.
128. "Father of U.S. Jet Flier Scoffs at Peiping Report," Shamokin, PA *News-Dispatch*, January 29, 1951, 2.
129. Sino Reds Give Propaganda A New Twist As Trap," *The Daily American*, Somerset, PA January 30, 1951, 2.
130. "Parents Get First Letter From Prisoner of War," *The Harrisburg Evening News*, August 2, 1951, 2.

there was never enough to heat huts and boil water. There was no medical care.[131] Two dozen men died each day by the end of February.[132]

Men died by the hundreds from injuries they suffered being tortured, from disease, and from the effects of malnourishment. Every morning men woke up next to dead bodies. Those who awoke to live another day stripped away any clothing that might offer the living some warmth. Burial details carried the dead up the hill to a resting place. The ground was too frozen to dig a grave, so the bodies were covered with rocks and ice. A man from the deceased's platoon would say a few words and move on.[133]

When it was his turn to carry off a dead American, Philip, like Rubin, would recite the Jewish Rachamim in Hebrew then in English.[134]

> *O God, full of compassion, Thou dwellest on high! Grant perfect rest beneath the sheltering wings of Thy presence, among the holy and pure who shine as the brightness of the firmament unto the soul of _____ who has gone into eternity. Lord of mercy, bring him under the cover of Thy wings, and let his soul be bound up in the bond of eternal life. Be thou his possession, and may his repose be peace. Amen.*[135]

---

131. William Clark Latham Jr. *Cold Days in Hell: American POWs in Korea*, (College Station, TX: Texas A&M University Press, 2012), 131.
132. William Clark Latham Jr. *Cold Days in Hell: American POWs in Korea*, (College Station, TX: Texas A&M University Press, 2012), 132.
133. William Clark Latham Jr., *Cold Days in Hell: American POWs in Korea*, (College Station, TX: Texas A&M University Press, 2012), 129-133.
134. Daniel M. Cohen, *Single-Handed: The Inspiring True Story of Tibor "Teddy" Rubin – Holocaust Survivor, Korean War Hero, and Medal of Honor Recipient*, (New York, NY: Berkley Caliber, 2015), 214-215.
135. Charles Rosenthal et. al., Prayers and Meditations (New York, NY: Riverside Memorial Chapel, 1964) 8.

Philip survived bouts of hepatitis, dysentery, and pneumonia.[136] After months of eating terribly undercooked hard grains, several more of his teeth were broken. A captured American doctor, Captain William Shadish, used primitive instruments with no anesthesia to pull damaged and infected teeth, including several of Philip's.[137]

Starting in March 1951, the Chinese conducted constant interrogations, incessant psychological warfare, round-the-clock indoctrination, and endless instruction in communist doctrine. The communists offered promises of better treatment if they cooperated. Philip knew better.

The men had to attend daily "reeducation" classes outdoors while temperatures hovered around freezing. Classes lasted from two to six hours. Chinese instructors, speaking mostly in broken English, lectured them about how they had been exploited by imperialism and that communism would free them from the yoke of capitalist oppression. They were told that the faster they learned their lessons, the sooner they could return to their homes and families.[138]

Most of the prisoners did not succumb to such brainwashing.[139] Many actively resisted. The average POW, like Philip Aaronson, knew from his upbringing and education that what he was being forced to listen to was wholly false. Their captors constantly harassed and threatened them into writing letters they promised to send home, but only if the words met with the Chinese leadership's approval. In hopes of

---

136. 6004th Air Intel Sv Sq, *Air Intelligence Information Report*, "Subject Interrogation of a Returned USAF Prisoner of War – Philipp (NMI) Aaronson, S/Sgt, USAF," Sep 53, p10. NARA RG 341 Series P 268-A Box 1 of Loc 170-63-26-06 Folder: Aaronson, Philip.
137. William Clark Latham Jr. *Cold Days in Hell: American POWs in Korea* (College Station, TX: Texas A&M University Press, 2012), 142.
138. Raymond B. Lech, *Broken Soldiers*, (Urbana, IL: University of Illinois Press, 2000), 92-94.
139. William Clark Latham Jr. *Cold Days in Hell: American POWs in Korea*, (College Station, TX: Texas A&M University Press, 2012), 197-199.

getting word out that they were alive, most captives eventually found ways to get the censors' approval of their carefully chosen words.

In March 1951, the Chinese published the two letters attributed to Staff Sergeant Philip Aaronson in one of their propaganda pamphlets. It was a small six-page booklet announced in a Peking newspaper with a photo of Philip among five fellow POWs.[140] The pamphlet contained the text of the fabricated letters. On the outside covers were horrific pictures of a dead GI and a dead Korean mother and child.[141] The Chinese widely circulated the pamphlet among communist countries and Marxist groups around the globe.

By spring 1951, well over a thousand U.S. POWs had died at Camp 5. Although he was starved, beaten, ravaged by diarrhea, had most of his teeth broken, and weighed less than a hundred pounds, Philip Aaronson survived. His body was broken but his fighting spirit persisted. Along with many others, he began to pursue a strategy of active resistance. He did not openly defy the communists, as did Major MacGhee, a leader of the so-called "resisters."[142] Philip's resistance was not physical. It was cognitive.

Behavioral Scientist Albert D. Biderman identified this form of resistance in his epic work for the Air Force, publicized in his book *March to Calumny*[143]. Biderman based his work on extensive interviews and assessment of documents as project scientist for an Air Force Korean War POW study. In his analysis of repatriated Korean War POWs, Biderman challenged the then-prevailing notion that the

---

140. "Somewhere in Korea," *The Morning Call*, Allentown, PA, March 24, 1951. 6.
141. SGM Herbert A. Frieman (Ret.), "Communist Korean War Leaflets," https://psywarrior.com/NKoreaH.html
142. Affidavit of Major David F. MacGhee, subscribed and sworn to by Major Frank M. Finn, JAGC, 13 April 1954. http://www.kpows.com/amazingveterans.html. Army Security Center 8589th AAU Fort George G. Meade, Maryland, Interview Report, Mac Ghee, David F. ASCIR #0067 28 June 1954.
143. Albert D. Biderman, *March to Calumny: The Story of American POWs in the Korean War* (New York, NY: The Macmillan Company, 1963), 58-61.

POWs, especially the senior officers, failed to resist and were "brainwashed" by communist techniques of mind control. Instead, Biderman found:

> *Covert organizations led by low-ranking individuals who are least subject to pressure and surveillance have been mentioned as necessary for effective resistance to an exploitative captor...There is one kind of relevant experience that is common to most American males, however. This is the frequent defiance, evasion, and harassment of school authorities by students.*[144]

One stalwart leader of this cognitive campaign was Father Emil Kapaun, a regimental Chaplain in the 1st Cavalry Division. He violated Camp 5 rules by regularly sneaking out almost every night to visit the enlisted men's compounds, bringing words of encouragement and food he had stolen from the communist's own supplies. Father Kapaun was ready at strategic moments with subtle rebuttals to communist indoctrination. After one three-hour lecture by Chinese Major Comrade Sun, Father Kapaun meekly asked questions that implied refutation of each tenet of Marx and Mao. When Major Sun said before the group that Father Kapaun was preaching propaganda, the chaplain matter-of-factly replied that he spoke only of "…Christian love, and I shall pray for your soul." He disarmed his enemies intellectually and spiritually. [145]

As word of his courage spread, some of the more self-confident lecturers challenged Father Kapaun's faith in God. One Korean War

---

144. Albert D. Biderman, "Cultural Models of Captivity Relationships," AFOSR-452 (Washington, DC: Bureau of Social Science Research, Inc, February 1961), 19-20. BSSR Research Report 339-4, Contract No AF 49(638)727, Behavioral Sciences Division, Air Force Office of Scientific Research. https://apps.dtic.mil/sti/tr/pdf/AD0257325.pdf

145. William L. Maher, *A Shepherd in Combat Boots: Chaplain Emil Kapaun of the 1st Cavalry Division* (Shippensburg, PA: Burd Street Press, 1997), 134-135.

survivor recalled an exchange between the chaplain and a Chinese instructor: "Where is your God now? Let him come and take you from here. See if he can feed you. You should thank Mao Tse Tung and Stalin for your daily bread. You cannot see or feel your God; therefore, he does not exist."

The communist had fallen into the conundrum that confronts nonbelievers. Father Kapaun rebutted his accusers like Job dealing with Eliphaz, Bildad, and Zophar.

> *"God is as real as the air that you breathe but cannot see," he said. "One day the Lord will save the Chinese and free them from the scourge that has set upon them. The Good Lord, as He fed thousands on the mountain, will take care of us. Mao Tse Tung could not make a tree or a flower or stop the thunder and lightning."*

With that, the humbled instructor gave up the argument.[146]

The Chinese soon came to fear Father Kapaun's informal power and influence. They stripped him of his rosary beads, but he fabricated another out of barbed wire. He continued to circulate among the enlisted prisoners under cover of darkness to share brief clandestine prayers with them. His support immeasurably sustained these desperate men, Protestant, Jewish, Catholic, and even professed atheists.[147]

To deal with this growing threat to their authority, the Chinese tortured two officers into accusing Father Kapaun of slandering the Chinese and threatening to court-martial POWs who cooperated with their enemy. Yet, nothing came of it, for the Chinese feared

---

146. William L. Maher, *A Shepherd in Combat Boots: Chaplain Emil Kapaun of the 1st Cavalry Division* (Shippensburg, PA: Burd Street Press, 1997), 135.
147. William Clark Latham Jr. *Cold Days in Hell: American POWs in Korea* (College Station, TX: Texas A&M University Press, 2012), 143.

widespread mutiny against them if they did anything obvious to Father Kapaun.[148]

Father Kapaun led a moving Easter Service on March 25, 1951. It had a profound effect on American prisoners, motivating them to press on despite the starvation, disease, torture, and death that surrounded them. The communists dared not attempt to impede the event as POWs lifted their voices in hymns and patriotic songs. But within a few months, Father Kapaun's own suffering and labor took their toll on his emaciated body. The Chinese hastened his death by withholding food and medicine needed to restore him to health. He died a martyr's death on May 23, 1951, with the Chinese, incredibly, attributing cause of death to syphilis.[149]

In June a Chinese broadcaster again read Philip's alleged letters over Radio Peking. On receiving word, Philip's mother said she "hopes and prays" that he'll be released if the truce brings about an exchange of prisoners, and she "would be the happiest mother on earth." She was grateful he was alive and added that she "prayed that everything will work out all right for the good of the whole world."[150]

Meanwhile, Philip was making his opening gambit in the cognitive resistance campaign as the Communists tried to lure their captives into joining a "peace" movement.[151] A new Chinese Commandant, Lieutenant Colonel Ding Chan,[152] came up with the idea. The POWs called him "Snake Eyes" because of his vulture-like face. In January

---

148. William L. Maher, A Shepherd in Combat Boots: Chaplain Emil Kapaun of the 1st Cavalry Division, (Shippensburg, PA: Burd Street Press, 1997), 137. See also *The Bible*, Matthew 21:45-46.
149. William L. Maher, *A Shepherd in Combat Boots: Chaplain Emil Kapaun of the 1st Cavalry Division* (Shippensburg, PA: Burd Street Press, 1997), 152.
150. "Mother Praying Son Be Freed," *The Harrisburg Evening News*, July 3, 1951, 1.
151. Raymond B. Lech, *Broken Soldiers*, (Urbana, IL: University of Illinois Press, 2000), 99-103.
152. Army Security Center 8589th AAU Fort George G. Meade, Maryland, Interview Report, Mac Ghee, David F. ASCIR #0067 28 June 1954, 36.

1951, the communists had tried to get prisoners held at the Kangye Camp, Camp 10, to sign a letter supporting the Stockholm Peace Appeal drafted by the communist World Peace Committee prior to the Korean War. That document called on the United Nations to outlaw nuclear war. It seemed innocuous at the time, but the Chinese added language to it and told the POW leaders to draft their own such letter. Few POWs complied. Many of those that did deliberately misspelled their names or signed in scribble impossible to read.[153]

Learning from their Kangye failure, the Chinese applied those lessons to their attempt at Pyoktong. They began by focusing on officers they believed would be most susceptible to signing the Stockholm Peace Appeal because they were in the worst physical condition. The communists had all officer POWs debate the matter for three days straight. Gradually, more and more of them came around to the "yes" group. The last of the holdouts capitulated on the third day when the senior POW officer told them that he would take responsibility for all.

Snake Eyes Ding grinned and said, "By accepting our viewpoint, you have joined the camp of peace. We will no longer treat you as war criminals but as liberated officers. Since you have joined the camp of peace, you must now translate your thoughts into deeds."[154] He told them they would march in a parade through the streets of Pyoktong with signs and banners for the awaiting communist "journalists" to document the event. The Chinese promised they would have a banquet after the parade, serving up pork and corn fritters.[155] The Com-

---

153. William Clark Latham, Jr. *Cold Days in Hell: American POWs in Korea*, (College Station, TX: Texas A&M University Press, 2012), 128-129.
154. Maj. Ronald Alley, USA Transcripts, 7:725, 4:177-78, 4:180, 4:189, 6:397, 6:444-45; Erwin Transcripts, 1:142a; and Cpl Claude Batchelor, USA, Transcripts, 8:1879, quoted in Raymond B. Lech, *Broken Soldiers* (Urbana, IL: University of Illinois Press, 2000),101.
155. William Clark Latham, Jr. *Cold Days in Hell: American POWs in Korea* (College Station, TX: Texas A&M University Press, 2012), 191-192.

munists thought they had the prisoners psychologically where they wanted them.

Each company drew up a banner on white cloth with "WE WANT PEACE!" written in red paint. Every POW was given a pennant on a stick with the word "PEACE" written across it.[156] As the men marched into Pyoktong village they could see a gaggle of reporters with cameras lined up, apparently to document the event. Philip and his crewmen were in the lead unit of the Sergeants Company. The men carrying the banner in front of the formation held it upside down. The others followed their lead by holding their pennants upside down. Then the Chinese told them to shout, "We want peace!" loud enough for their tape recorders and movie cameras. Instead, they yelled, "We want meat!" The angered Chinese officers made them march past the reporters again. This time they dragged their pennants in the dirt and shouted, "We want rice!" Their "feast" that night was boiled squash.[157]

Ding then had his henchmen try to demoralize prisoners by forcing them to participate in a "rally for peace" inside the camp. He harangued them once more into signing a "peace appeal" letter addressed to "the Five Great Powers." This time they had to line up in front of a table with a copy of the Stockholm Peace Appeal and a letter of endorsement for them to sign. Initially the Americans resisted. The Communists argued the move would serve as "proof of life." Then they promised better food and living conditions to all men who would sign. Some historians conclude, based on opinions of some repatriated officers, that everyone did so.[158] But individual accounts by many

---

156. Raymond B. Lech, *Broken Soldiers* (Urbana, IL: University of Illinois Press, 2000), 101.
157. William H. Funchess, *Korea P.O.W.: A Thousand Days of Torment* (Clemson, SC: South Carolina Military Museum, n.d.), 79-80.
158. Raymond B. Lech, *Broken Soldiers* (Urbana, IL: University of Illinois Press, 2000), 102. William Clark Latham, Jr., *Cold Days in Hell: American POWs in Korea* (College Station, TX: Texas A&M University Press, 2012), 192.

enlisted men—and recent documents made available in the National Archives—prove that at least some, and probably many, did not.[159]

Staff Sergeant Philip Aaronson did not sign. He had shared with his B-29 crewmates, also imprisoned at Camp 5, how the Chinese had twisted and added to his words in the letter to his parents at the Pukchin interrogation center back in January. Likewise, Flight Engineer Tech Sergeant James Edwards, Radio Operator Staff Sergeant Jimmy Sanders, and Side Gunner Sergeant Victor Foote refused to sign. Tech Sergeant Edwards and Sergeant Foote survived the war and would corroborate Philip's account.

Conditions began to improve at Camp 5 in the summer of 1951 as armistice discussions got underway. Food portions reached survival levels and were more nutritious. Some medicines were available, and Philip recovered from his continuing dysentery, jaundice, pneumonia, and hepatitis. He turned in his tattered remnant of a uniform for clean underwear, blue uniforms, and canvas sneakers. Mess kits, barber supplies, and English language magazines began to appear.[160] They were Marxist journals, but some also carried western sports news that was less subject to propagandizing. In the warmer weather, he was allowed to bathe in the Yalu River.

Although Philip recovered from his sicknesses, his battle buddy, Radio Operator Staff Sergeant Jimmy Sanders, did not. He had contracted pellagra, a disease brought on by lack of vitamin B. A lean man, Jimmy Sanders had lost over half his body weight by that summer, and despite the improved chow, he was too far gone to recover. This disease causes reddened, thickened, cracked skin on the top surfaces of hands, feet, and chest, where 3-inch, puss-filled sores erupt.

---

159. Brett Fearer, "Prodigals or Traitors: American POWs During the Korean War, Brainwashing, and National Security" (master's thesis, SMU, 2023), 46, https://scholar.smu.edu/cgi/viewcontent.cgi?article=1025&context=hum_sci_history_etds

160. William H. Funchess, *Korea P.O.W.: A Thousand Days of Torment* (Clemson, SC: South Carolina Military Museum, n.d.), 75.

It is accompanied by diarrhea and dementia that lead to complete apathy for life. Untreated, pellagra almost inevitably leads to a slow, painful death. No treatments for pellagra were available at Pyoktong or any other POW Camp in North Korea.[161] Jimmy died on August 10, 1951. Philip Aaronson, James Edwards, Robert Burke, Lyle Dodd, Billy Foshee, and David MacGhee, the surviving enlisted crewmen of B-29 814, resolved to press on to their return to the United States of America.

Lectures intensified in the summer of 1951, focusing on the texts of speeches written by Engels, Marx, and Lenin on the superiority of communism. Required readings included *The Decline and Fall of American Capitalism*, *Communist Manifesto*, and *The Life of Karl Marx*.[162] Such political philosophies rang deafeningly silent to Philip Aaronson. He was steeped in the Law and the Prophets from boyhood and had memorized the Declaration of Independence and the Preamble to the U.S. Constitution in school. His Chinese indoctrinators had no comprehension of their own blind faith in the false religion of communism.

Toward the end of October 1951, the Chinese moved the officers and airmen at Camp 5 to another permanent camp eight miles away.[163] When they arrived, Philip figured he had marched a combined total of 800 miles across North Korea.[164]

---

161. William Shadish, M.D. with Lewis H. Carlson, *When Hell Froze Over: The Memoir of a Korean War Combat Physician Who Spent 1010 Days in a Communist Prison Camp* (New York: NY, iUniverse, inc., 2007), 135.
162. Raymond B. Lech, Broken Soldiers (Urbana, IL: University of Illinois Press, 2000). 94.
163. William Clark Latham, Jr., *Cold Days in Hell: American POWs in Korea* (College Station, TX: Texas A&M University Press, 2012), 203.
164. "Freed Airman Reports Reds Lied in Propaganda Releases," *The Wilmington Morning News*, September 29, 1953, 4.

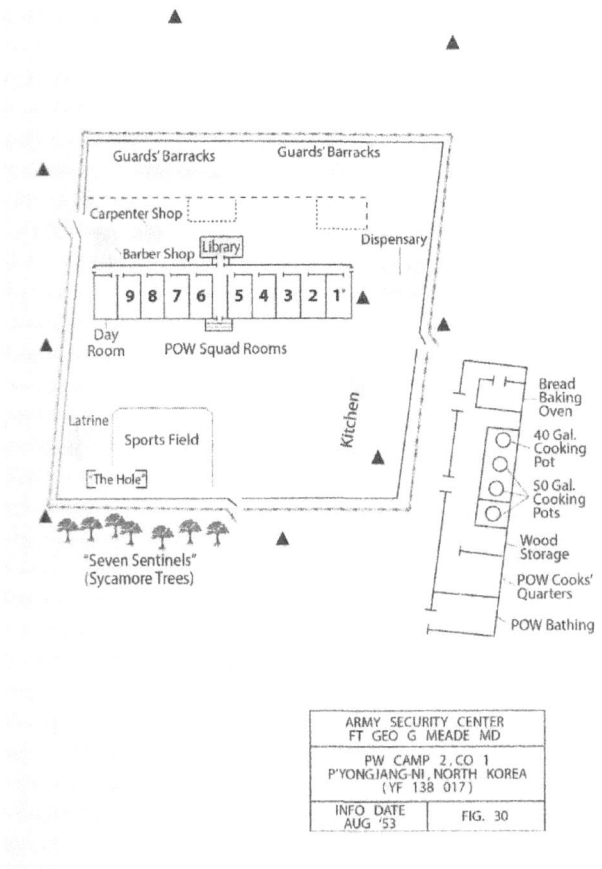

Camp 2 Pin-Chong-Ni, Prisoner Compound[165]

Camp 2 was marginally livable. Prisoners were housed, 24 men to a room, in a Japanese occupation-era schoolhouse that the Americans had to liberate from rodents, insects, and flies.[166] "In Your Face" reac-

---

165. Army Security Center 8589th AAU Fort George G. Meade, Maryland, Interview Report, Mac Ghee, David F. ASCIR #0067 28 June 1954.
166. William Shadish, M.D. with Lewis H. Carlson, *When Hell Froze Over: The Memoir of a Korean War Combat Physician Who Spent 1010 Days in a Communist Prison Camp* (New York: NY, iUniverse, inc., 2007), 66.

tionaries, such as Major MacGhee, were subjected to physical punishment during their stay. For the Chinese, that was in part an effort to get them to sign confessions for their "crimes." It was also retribution for their attitudes and actions of resistance and humiliation.[167] For the men of the cognitive resistance, Camp 2 was ideal terrain for their concept of operations.

At Panmunjom in December 1951, the Communists provided a roster of 2,724 U.S. POWs said to be alive in their camps. The Defense Department carefully correlated names and serial numbers, then published the list in print, arranged alphabetically.[168] Since Philip's family name began with the unusual double-letter "Aa," his was the first name on the list. In Harrisburg, Pennsylvania, months of suspense turned suddenly into apprehensive joy for the parents of eight, who learned that their son in the Air Force was alive, although a prisoner of war.

> *Our son's name was the first on the list," said Mrs. Dorothy Aaronson, whose son, "Sergt." Philip Aaronson, 25, was reported missing in action in Korea November 11, 1950. "Sergt." Aaronson, an aerial gunner, was reported a prisoner of war by the Peiping Radio as recently as last June. But the soldier's father, Hyman Aaronson, said recent reports of prisoner atrocities forced the family to all but give up hope for Philip's life. "We were under an awful suspense to begin with," said Mr. Aaronson. "But after we heard of those boys who were supposed to have been killed by the Communists, our morale went pretty low. Now we don't know whether we are coming or going, we're so happy." "Philip is our third child," said Mrs Aaron-*

---

167. Affidavit of Major David F. MacGhee, subscribed and sworn to by Major Frank M. Finn, JAGC, 13 April 1954. 17-18 http://www.kpows.com/amazingveterans.html
168. "List of U.S. Prisoners of War That Was handed to the U.N. Command by Communists," *New York Times*, December 20, 1951, 10.

son. "And he has five sisters and two brothers who are terribly happy right now."[169]

In February and March 1952, Communist officials accused the United States of conducting germ warfare on North Korea.[170] They presented evidence in the form of coerced "confessions" by tortured American airmen. In October, the Chinese presented a "scientific" study at the United Nations accusing the United States of waging germ warfare.[171] Sociologist Albert Biderman's post-war analysis for the Air Force showed conclusively that this propaganda had no significant impact on public opinion around the globe.[172]

In many ways the communists had given counsel to their own fears. An outbreak of smallpox in North Korea took a severe toll on the Korean and Chinese people. When radar-jamming foil "chaff" fell nearby, Chinese officers donned their protective suits to collect the pieces of tinsel to subject them to laboratory tests that, of course, "proved" their case.[173]

Philip Aaronson turned to his art to seize the moment. The schoolhouse barracks at Camp 2 provided luxury accommodations for wharf rats in the walls.[174] With a couple of compatriots, Philip captured

---

169. "Happiness Grips Kin of GIs; Many Had Abandoned Hope," *Washington Evening Star*, December 19, 1951, 20.
170. William Clark Latham, Jr., *Cold Days in Hell: American POWs in Korea*, (College Station, TX: Texas A&M University Press, 2012), 181.
171. Dan King, *The Yalu River Boys: The true Story of a B-29 Bomber Crew's Combat and Captivity in the Korean War* (North Charleston, NC: Pacific Press, 2018), 215-216.
172. Albert D. Biderman, *March to Calumny: The Story of American POWs in the Korean War*, (New York, NY: Macmillan, 1963), 77-78.
173. William Shadish, M.D. with Lewis H. Carlson, *When Hell Froze Over: The Memoir of a Korean War Combat Physician Who Spent 1010 Days in a Communist Prison Camp* (New York: NY, iUniverse, inc., 2007), 76. William H. Funchess, *Korea P.O.W.: A Thousand Days of Torment* (Clemson, SC: South Carolina Military Museum, n.d.), 115.
174. William H. Funchess, *Korea P.O.W.: A Thousand Days of Torment*, (Clemson, SC: South Carolina Military Museum, n.d.), 90.

one, shaped a parachute with "USAF" lettered on it, and tossed it out where the guards would find it.[175] The stunt terrorized the Chinese.[176]

An international commission toured alleged germ warfare sites and interviewed "witnesses" who claimed that U.S. aircraft had air-dropped "poisonous fleas, flies, mosquitos, rats, and clams."[177] Knowing these reports to be blatantly false, many U.S. airmen pretended to confess, relying on the expectation that their acknowledgements, many riddled with errors on simple facts, could not be corroborated and would be recognized by military officials as forced statements. Upon repatriation in 1953, they unanimously renounced their confessions and blamed them on communist torture. Post-war investigations of repatriated POWs exonerated them all of any misconduct on the matter of germ warfare confessions.[178]

Encouraged by the results of their germ warfare escapades, the cognitive resisters stepped up their efforts. One merry band dug a hole in the ground behind their building. They planted a sheet of paper at the bottom and re-filled the hole. Guards dug it back up. It read: "Mind Your Own Business." A particularly rude sentry would kick the sleeping POW's boots as he made his rounds at night. So, the men nailed a pair of boots to the floor and filled it with rocks.[179] Ouch!

They even exploited the Chinese' lack of understanding of American culture in their "confessions" and "self-criticisms."[180]

---

175. "Freed POW Tells of Germ Warfare," *The Harrisburg Patriot-News*, September 6, 1953, 2.
176. "Freed POW Tells of 'Germ Warfare'," *The Harrisburg Patriot-News*, September 6, 1953, 2.
177. William Clark Latham, Jr. *Cold Days in Hell: American POWs in Korea*, (College Station, TX: Texas A&M University Press, 2012), 182.
178. William Clark Latham, Jr. *Cold Days in Hell: American POWs in Korea*, (College Station, TX: Texas A&M University Press, 2012), 184, 228-230.
179. William Clark Latham, Jr. *Cold Days in Hell: American POWs in Korea*, (College Station, TX: Texas A&M University Press, 2012), 208.
180. Albert D. Biderman, *March to Calumny: The Story of American POWs in the Korean War*, (New York, NY: The Macmillan Company, 1963), 57-58.

- *I promise never again to call Wong, that no-good, dirty, son-of-a-b***.*
- *I was blinded by the Wall Street imperialists. Because I was taking home $15 a week.*
- *One of the questions was "Who is your best friend?" I put "The only friend I have wants to borrow money."*
- *Q: "Who was your commander? A: "Military secret, can't be given."*
- *Answers to a question on a form:*
  - *"I filled out something like this before and I ended up in the Army."*
  - *"This is too much for me. I request transportation to go back and have my lawyer fill it out."*

The triumph of the cognitive resisters of Pin-Chong-Ni was "crazy week." Men spoke gibberish, walked imaginary dogs, performed Native American rain dances, and flew about the yard like combat jets while 'strafing' the guards.[181] Others,

> ...rode an imaginary horse, galloping to-and-fro, brushing and grooming his steed. Another man cut his hair into a Mohawk, wrapped a blanket around his shoulders and performed a native-American war dance. Capt Nardella adopted an imaginary dog that he taught to do several tricks. However, the mutt had a bad habit of 'urinating' on the legs of any Chinese that made the mistake of getting too close. Nardela would scold the invisible puppy and sincerely apologize to the guards whose trousers his pet had soiled.[182]

---

181. William Clark Latham, Jr. *Cold Days in Hell: American POWs in Korea* (College Station, TX: Texas A&M University Press), 2012. 208.
182. Dan King, *The Yalu River Boys: The True Story of a B-29 Bomber Crew's Combat and Captivity in the Korean War*, (North Charleston, NC: Pacific Press, 2018), 218.

The coup de maître was delivered by Navy Lieutenant John "Rotorhead" Thornton who "…spent the entire week wearing a beanie cap outfitted with propellers and pretending to ride an imaginary motorcycle."[183]

Lieutenant Thornton's POW story is remarkable. His helicopter was shot down while he was attempting to extract a behind-the-lines agent who had critical intelligence needed for an immediate strike by naval gunfire and attack fighters. Lieutenant Thornton and the agent evaded captivity for 10 days. Moving stealthily through wooded mountainous terrain toward friendly lines, he covered 50 miles but was caught just 18 kilometers from the Americans.[184]

Throughout his captivity, Lieutenant Thornton determinedly resisted physical and psychological torture in the vilest conditions. It was 307th Wing B-29 pilot and fellow POW William C. McTaggart, Jr. who awarded him the callsign "Rotorhead" while both were imprisoned at the infamous interrogation center known as "Pak's Palace." Lieutenant Thornton earned the callsign because, according to McTaggart, he "was always in a rotating configuration while spinning off to the crap hole." [185]

The POWs knew that the Chinese were censoring their mail, both sent and received. The captured doctor, Captain William Shadish, tells of an interpreter who told him, "The government had your wife and child pose with a new car in Washington to propagandize how good life is in America." The doctor told the interpreter that he and his wife own that car. But the interpreter insisted, "You could not own that car." He believed no common person could own his own automobile.[186]

---

183. William Clark Latham, Jr. *Cold Days in Hell: American POWs in Korea* (College Station, TX: Texas A&M University Press, 2012), 208.
184. Captain John W. Thornton with John W. Thornton, Jr., *Believed To Be Alive* (Middlebury, VT: Paul S. Eriksson, 1981) 82-103.
185. Captain John W. Thornton with John W. Thornton, Jr., *Believed To Be Alive* (Middlebury, VT: Paul S. Eriksson, 1981) 142.
186. William Shadish, M.D. with Lewis H. Carlson, *When Hell Froze Over: The Memoir of a Korean War Combat Physician Who Spent 1010 Days in a Communist*

On April 5, 1953, Philip wrote a letter to his parents that got through to them in June. He told them he had just returned from Easter services and that his volleyball team made it to the camp tournament semifinals. He wrote that he was now allowed to write as often as he wanted and the letters could be long. Telling them of his future plans on return to build a home and buy a car, he asked them to send him pictures of "new model automobiles."[187] He wanted to drive home to his Chinese keepers the insight gained from Dr. Shadish's conversation with the censor.

In the summer of 1952, Philip became a ringleader among the highly effective cognitive resisters among U.S. POWs in the Korean War, the group Biderman identified as "The Show." This faction often mocked communist doctrines and personalities in ways largely oblivious to their Chinese and North Korean captors, but quite evident to soldiers and airmen steeped in American culture and military experience.[188]

Philip's high school stardom acting in school plays, announcing for school programs, and serving as master-of-ceremonies for assemblies, had grown in his early Army Air Forces Days into a popular side-gig in the service's entertainment programs at bases, stations, and camps. As the Communists began to allow POW holiday celebrations with skits, plays, and shows, Philip emerged as a natural leader.

In June the theatrical committee put on a performance of *Hells-a-Done Popped*, a satirical take-off on *Hellzapoppin*.[189] Props and costumes were creative masterpieces fabricated by British POWs from scraps and detritus gathered from around camp. In the Camp 2 rendi-

---

Prison Camp (New York: NY, iUniverse, inc., 2007), 70.
187. "Family Receives POW's Letter," *The Harrisburg Evening News*, June 13, 1953, 1.
188. Albert D. Biderman, *March to Calumny: The Story of American POWs in the Korean War* (New York, NY: The Macmillan Company), 57 – 59.
189. IMDb, "Hellzapoppin," 1941. https://www.imdb.com/title/tt0033704/

tion of the closing sketch, the cast gets their first look at the Smithsonian's *Peking Man*. Chinese officers' faces turned from laughter to rage as the tour guide's dialogue was translated to reveal that the Chinese had not evolved after hundreds of thousands of years. The insult resulted in isolation sentences for participants and a month long ban on theatrical performances.[190]

For Christmas 1952, Snake Eyes Ding allowed the men to put on a play, though it had to be pre-approved before Ding allowed them to perform it. They seemed to have chosen the 1935 C.S. Forrester novel, *The African Queen*, as the basis for their script, the book that in 1951 was turned into the Academy Award winning motion picture.[191] The Chinese censors perhaps thought the World War I plot pitting Britain versus Germany would be harmless. But the cognitive resisters turned it into a representation of the United Nations versus the Chinese in the present war, and the resilience of the American spirit to rise up and conquer all.

"I don't know why the Germans would want this God-forsaken place."

"Nothin' a man can't do if he sets his mind to it. Never say die. That's my motto!"

"Dear Lord, we've come to the end of our journey, and in a little while we'll stand before you. I pray for you to be merciful. Judge us

---

190. Dan King, *The Yalu River Boys: The True Story of a B-29 Bomber Crew's Combat and Captivity in the Korean War*, (North Charleston, NC: Pacific Press, 2018), 212.
191. Dan King, *The Yalu River Boys: The True Story of a B-29 Bomber Crew's Combat and Captivity in the Korean War*, (North Charleston, NC: Pacific Press, 2018), 224. King asserts that the 1952 POW Christmas play was based on a Bogart and Sinatra movie titled *Whiskey River*. It may be that the play put on by the POWs was based on the popularity of the 1951 Bogart-Macall motion picture *The African Queen* that was adapted from the C.S. Forester novel of the same title.

not for our weaknesses, but for our love and open the doors of heaven for Charlie and me."[192]

At other times, musical performances included patriotic songs, such as "America the Beautiful," and even "The Star-Spangled Banner," numbers that provoked the ire of the communists. But they dared not interrupt for fear of inciting the prisoners to riot.

The final operation in the culture war occurred in May 1953, a few days after wounded prisoners were repatriated in Operation Little Switch. British POWs put on an adaptation of *Pygmalion*; the complex George Bernard Shaw play adapted for stage in 1914 from a classic Greek poem by Ovid. The American POWs were politely appreciative; and then came the traditional post-performance celebration with a discrete presentation of a cake that had somehow been baked with scarce ingredients, like sugar and flour, and decorated with icing in the form of the Stars and Stripes. The POWs of the two allied nations celebrated with a unison recital of the Pledge of Allegiance.[193]

The communists returned Staff Sergeant Philip (No Middle Initial) Aaronson to U.S. military custody at Panmunjom, Korea, on Saturday, September 5, 1953.[194] It was his 28th birthday. Fittingly, it was also, for Jews, Shabbat, a day of rest commemorating the liberation of the children of Abraham from slavery under the Egyptian pharaoh and the promise of a future perfect world.

A Jewish Chaplain was the first American to interview Philip at Freedom Village. The Red Cross set up a communications station for repatriates to send home a telegram, "Thank God I have been set free. Am well except for dental cavities." Medical officials sent a second

---

192. IMDb, "The African Queen Quotes," https://www.imdb.com/title/tt0043265/quotes/
193. Dan King, *The Yalu River Boys: The True Story of a B-29 Bomber Crew's Combat and Captivity in the Korean War*, (North Charleston, NC: Pacific Press, 2018), 229.
194. "Released American Prisoners," *Fort Worth Star-Telegram*, September 5, 1953.

message from a military hospital in Tokyo indicating he was treated there for "hypertensive vascular disease and dental cavities." His sister Henrietta told reporters, "Thank God. This is just wonderful. That's our Phil—God bless him."[195]

Dorothy Aaronson's sense of relief was deep. "I still have vivid memories of the horror of prisoners of World War I," she said. Her father left their home in Austria for America in 1910. By 1914 he had saved enough money to buy his family passage to join him. But the outbreak of the war interrupted their journey and resulted in her mother's death from malnutrition. Dorothy was among the first to donate blood through the Red Cross during the Korean War. "That is why I worried so much about my son. But now he will be home soon. It's just wonderful. I have so much to be thankful for."[196]

The Navy transported him on U.S. Navy Ship *General R.L. Howe* to Japan for medical evaluation and treatment. The ship then proceeded to San Francisco. During the Pacific voyage, repatriates were subjected to U.S. military intelligence interviews, additional medical care, and psychological interviews before being released.

Passing under the Golden Gate Bridge, he arrived in San Francisco on Tuesday, November 24th. The first thing he did was make a long-distance call to his parents, telling them he would arrive home that weekend. The Air Force gave him and fellow Harrisburg repatriate Charles B. Schlichter a three-day pass in the San Francisco area to visit his sister, married to now-Major Robert Cohen, U.S. Army, who also was a Korean War veteran.

Philip flew on a non-stop cross-country flight in a state-of-the-art Constellation airliner to Washington, D.C., where he got more work done on his teeth and donned his first pair of eyeglasses. The Air Force

---

195. "Red Cross Brings Cheering Word From Tokyo Hospital," *The Harrisburg Evening News*, September 21, 1953, 21.
196. "Photo of Released GI Son Delights Uptown Family," *The Harrisburg Evening News*, September 17, 1953, 21.

gave him a total of four months off between convalescent leave and accumulated furlough time. And he drew his back pay: "That should make a good picture – pushing a wheelbarrow away from the paymaster's window," he said.[197]

Philip chose not to reenlist when his term was up. He married hometown girl Lottie Gross in 1954. Back home in Harrisburg, he was often honored by various civic groups. He was always ready to extol the values of America. This B-29 gunner's message reverberates for us today.

I have seen little children 5 to 12 years old put into fields to work from dawn til dusk. Old women and mothers repairing roads and working in fields for no pay — their obligation to Communist society. Every able-bodied man conscripted into the Communist Army, those who refused exiled to labor camp or executed.

I have watched people where religion was no religion at all, only a program of Communist ideology and doctrine. All this, and much more, I saw in North Korea during 34 months as a prisoner of war held by North Korean and Chinese Communist forces. This is the life that people have to look forward to behind the Iron Curtain.

Now safely back in the United States of America, I am free to worship as I please, free to live where I choose, free to do the work I desire. I am happy and grateful to live among free people who enjoy individual dignity.

God bless our wonderful country, protect it, and keep us free.[198]

Philip Aaronson died on March 29, 1998, and was interred at Indiantown Gap National Cemetery, Pennsylvania.[199]

---

197. "Main Interest of Freed City POW Is Just 'Being Home," *The Harrisburg Patriot-News*, September 28, 1953, 1.
198. "What American Free Enterprise Means to Me," *The Harrisburg Evening News*, March 2, 1955, 8.
199. US Department of Veterans Affairs, National Cemetery Administration, Na-

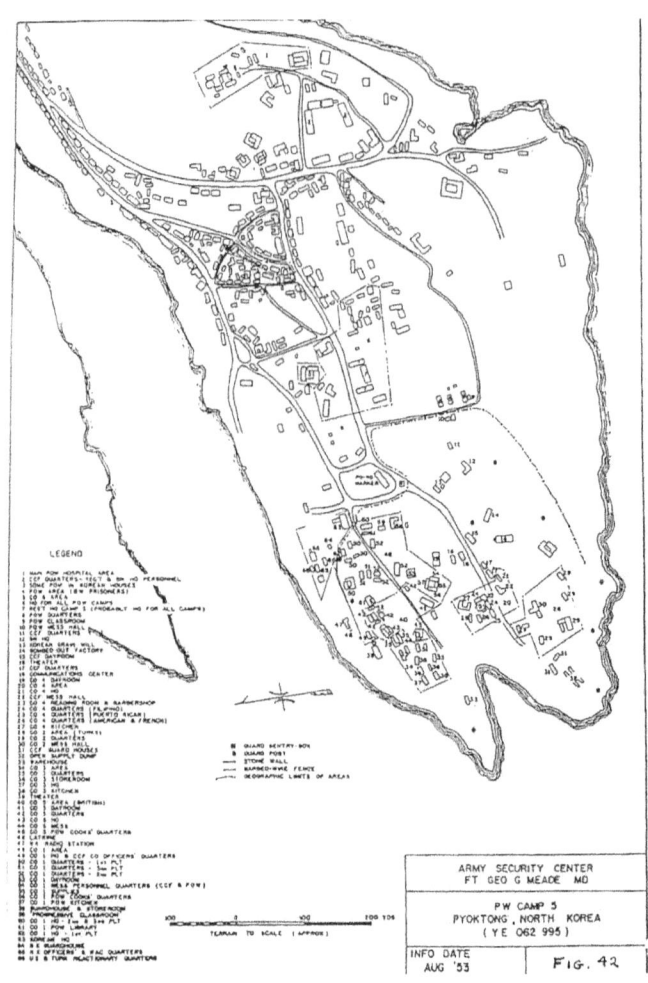

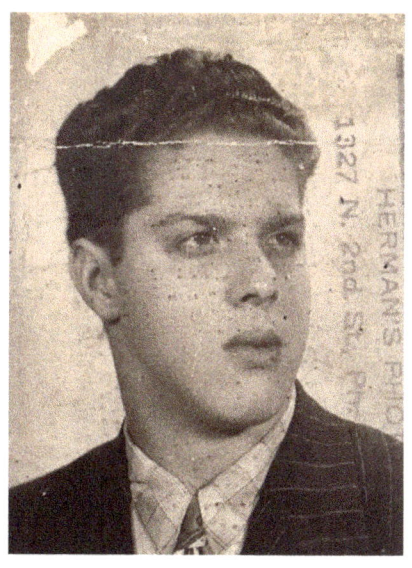

Philip Aaronson High School Portrait. (Source: Public Photos, Ancestry. com https://www.ancestry.com/mediaui-viewer/collection/1030/ tree/101265285/person/350088968900/media/6b9d1ee6-b8e1- 423d-be95-1728ad22677a?queryId=57899f82-24bb-4317-8cfe- 2550e5ff82cb&_phsrc=FEh73&_phstart=successSource)

Pvt Philip Aaronson on graduation from mechanic-gunner's course during World War II. (SOURCE: The Evening News, Harrisburg, Pennsylvania, June 6, 1944. https://www.newspapers.com/image/59638646/?match=1&terms=philip%20aaronson)

POWs march along a frozen trail during the winter of 1950. More than three thousand Americans were captured during the fighting in November and December...only half of the men captured during this period survived captivity. (Associated Press photo from William Clark Latham Jr., Cold Days in Hell: American POWs in Korea, (College Station, TX: Texas A&M University Press, 2012), 46)

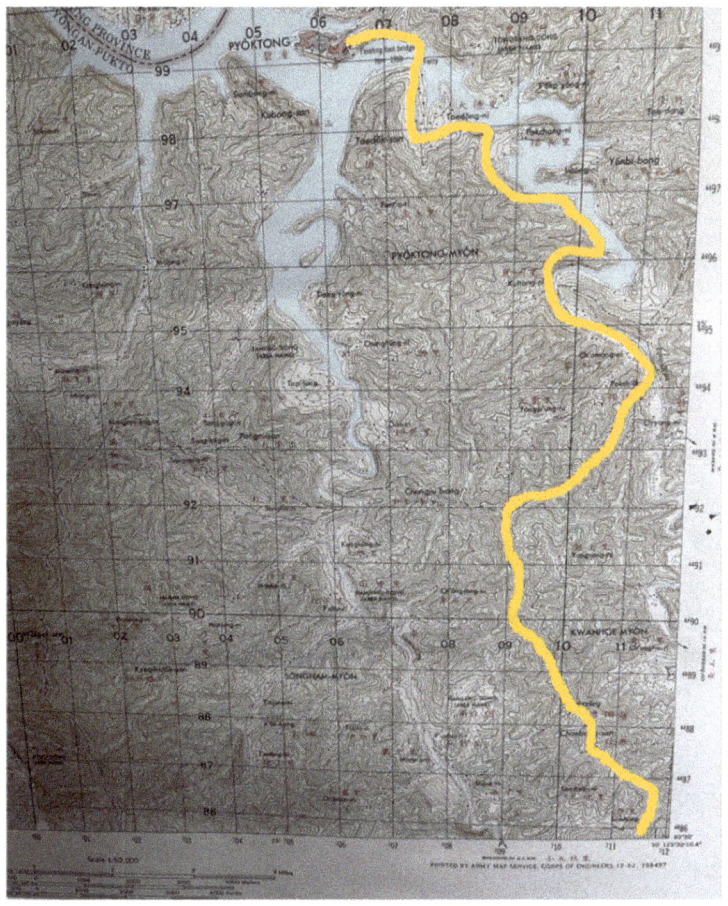

Author's reconstruction of plausible final 10 kilometers of the POW route to Pyoktong POW Camp, January 22 – 27, 1951. SSgt Philip Aaronson, USAF, was detailed to lead the 500+ enlisted men on this deadly, wind-swept march in temperatures ranging between -30° and -10° F. (Map: Pyoktong Korea, 1: 50,000. Sheet 62351 AMS Series L751, Third Edition. Army Map Service, Corps of Engineers U.S. Army, Washington, D.C., 1951, printed 12-52, 758497. NARA, RG 77, Army Map Series)

North Korean Families in Dugout Shelters. (Photos: Russian State Film and Photo Archive, Taewoo Kim, "Overturned Time and Space: Drastic Changes in the Daily Lives of North Koreans During the Korean War," Asian Journal of Peacebuilding Vol 2. No. 2 (2014), 246-247)

Front and Back Cover of communist propaganda pamphlet distributing fabricated letters coerced from S/Sgt Philip Aaronson. (Source: SGM Herbert A. Friedman (Ret.) "Communist Korean War Leaflets," https://www.psywarrior.com/NKoreaH.html)

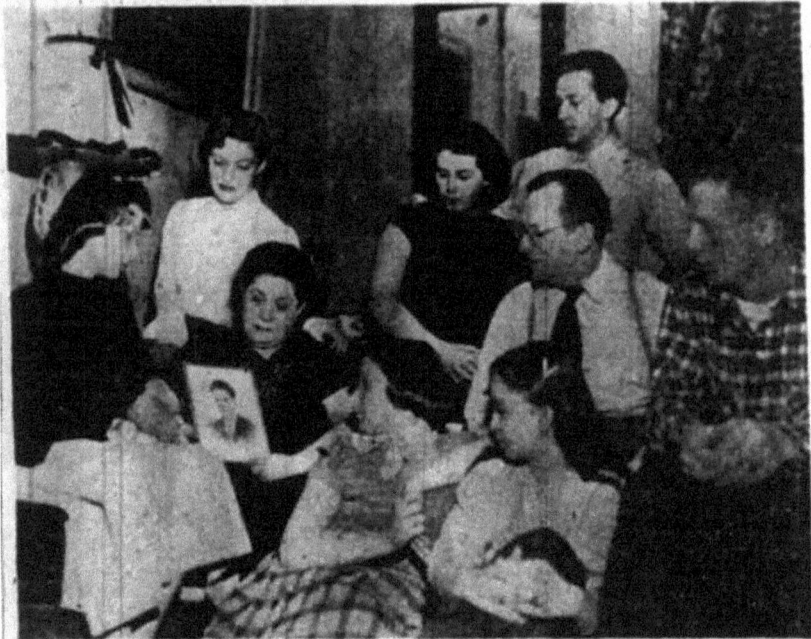

The South Bend Tribune, December 19, 1951,1.

Cowboys, soldiers, "girls" and men-about-town come onstage in the grand finale of a camp variety show.

An example of "The Show," Sociologist Albert Biderman's classification of POW resisters who employed humor, dramatic irony, turn of phrase, to rebut and refute communist propaganda and indoctrination. (United Nations' P.O.W.'S In Korea, Chinese People's Committee for World Peace, Peking, China, 1953, p. 46, https://archive.org/details/united-nations-pows-in-korea/page/46/mode/2up)

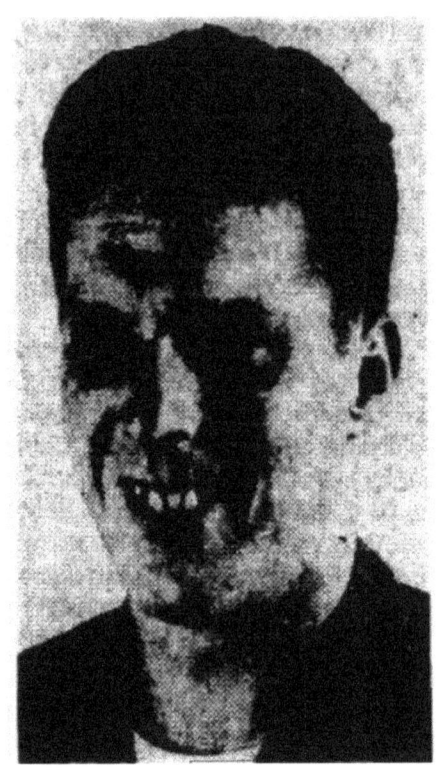

S/Sgt Aaronson, recovering at U.S. military hospital in Tokyo in photo released by the Red Cross. (Photo: "Red Cross Brings Cheering Word From Tokyo Hospital," The Harrisburg Evening News, September 17, 1953. 21.)

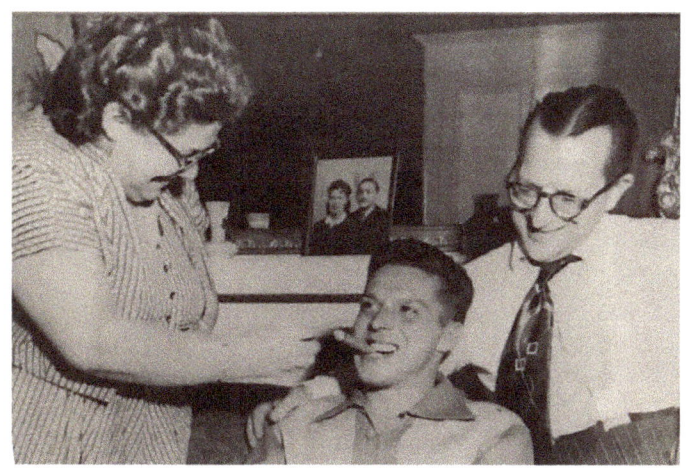

S/Sgt Aaronson celebrates his homecoming with his parents in late 1953 after surviving 34 months' captivity in North Korean POW camps. (Source: Public Member Photos, Ancestry.com https://www.ancestry.com/mediaui-viewer/collection/1030/tree/101265285/person/350088968900/media/9731af95-3a0c-477a-9d04-6d4f3cb6a4c4?queryId=69dc88e8-b22d-4c9f-9ccf-e73a004bf9fb&_phsrc=FEh64&_phstart=successSource)

Philip and Lottie Aaronson gravestone at Indiantown Gap National Cemetery, Pennsylvania. Source: https://www.cem.va.gov/nationwide-gravesite-locator/

CHAPTER FOUR
Dale H. Crist

"There d*** sure is a bullet out there that's got your name on it."[200]

U.S. Air Force Technical Sergeant Dale Crist was a highly trained and experienced gunner who completed 42 combat missions over North Korea. The Air Force awarded him the Distinguished Flying Cross, the Purple Heart, and two Air Medals for his combat service in the Korean War. He recorded his story for the Library of Congress Oral History Program.[201]

Dale Crist was born in Holly, a small town in southeast Colorado, and raised on a farm in Coolidge, Kansas, a half-dozen miles across the border from his birthplace. The family endured the Great Depression because they had a productive farm operation, with livestock, grain, and water sufficient to last to the end of the Dust Bowl years. Dale vividly recalled the big, rolling black cloud that made everything dark in the middle of the day and blasted dust everywhere.

He graduated from high school in 1946 and took jobs working on farms in the area. He entered a two-year Agricultural Degree program at Kansas State College of Agriculture and Applied Science, in Man-

---

200. Dale H. Crist Collection (AFC/2001/001/84581), Veterans History Project, American Folklife Center, Library of Congress. 53:05. https://www.loc.gov/item/afc2001001.84581/
201. Dale H. Crist Collection (AFC/2001/001/84581), Veterans History Project, American Folklife Center, Library of Congress. https://www.loc.gov/item/afc2001001.84581/

hattan, Kansas. In those days, all men in a land-grant college such as K-State were required to join the Reserve Officer Training Corps (ROTC). "That was the best thing I ever done," Dale said.

He could have been deferred from the draft because of his status as a college student, but "I didn't want to be a draft dodger, so I joined the Air Force," Dale said. "Ever since I was in ROTC, I wanted to be an aerial gunner, but the recruiter told me there was no chance because there was no gunnery school at that time."

He went to Lackland Air Force Base for Basic Military Training, hoping something would come up. Two weeks before graduation from Basic, the Air Force opened an aerial gunnery school in Denver to meet the demand for gunners in the Korean War. Dale, one of the top two graduates in all of Basic, was selected for Gunnery School. All 18 men in his gunnery class would go into combat in Korea. Only four lived through the war to come home.

Dale went from Gunnery School to Randolph Air Force Base, Texas, where he was assigned to a B-29 crew under Aircraft Commander Captain Badgett preparing for combat missions. In February of 1951, after eight months of training, the Air Force selected this crew for classified mission training in atomic bomb operations. Dale was one of three members of this "Lead Select Crew" who went to Chanute Air Force Base, Illinois, to master the specialized skills in how to employ an atomic bomb.

One practice mission took him over the Gulf of Mexico with a "dummy" bomb. It was a clear day when the bomb was dropped from 30,000 feet. "I was flying right gunner that day and followed that thing all the way to the ground," he said. "They had a big circle there, and the d*** thing hit in that circle." The atomic bomb couldn't be armed until the aircraft was airborne, and once armed, it couldn't be disarmed. Though the training flights used dummy bombs, they did contain explosives. "You had to know what you were going to do when you did it," Dale said. "You had to go out into the bomb bay in order to arm that thing after you were airborne."[202]

It took Captain Badgett and his crew 19 days and four engine changes to get from California to Okinawa. At Kadena, the 307th still lived in tent city. Crews billeted together, the six officers in one tent and the six enlisted in another close by. The mess hall, showers, and latrines were all in separate tents. Dale says the water in tent city was smelly and oily because it was shipped to the island on barges. The crew shared common miseries, but enlisted men of a combat crew pulled no KP or guard duty. And once aboard the aircraft, Dale said, "There was no rank among the crew. You trusted each other. You had a job to do, and you did it the best you could."[203]

Dale claims his B-29 was the only one that never aborted a flight. However, after "bombs away" on one mission, a red light in the aft gunners' compartment indicated that some bombs had not dropped. Dale had to go into the bomb bay to investigate. To his dismay, he saw that the top bomb in the rack was hanging from one of the two brackets. Worse, he saw that the fuze prop was turning, meaning the bomb was armed and could not be disarmed. One false move and the bomb could detonate inside the ship with fatal consequences for all aboard. Dale knew how the bomb bracket worked, so he borrowed a long screwdriver from the flight engineer and the radio operator's back-mounted parachute while the pilot descended to 10,000 feet, and the bombardier opened the bomb bay doors.

Dale reached in with the screwdriver and placed it so that it would push the bomb bracket to a release position, whereupon the bombar-

---

202. Dale H. Crist Collection (AFC/2001/001/84581), Veterans History Project, American Folklife Center, Library of Congress. https://www.loc.gov/item/afc2001001.84581/ 25:50 – 26:00.
203. Dale H. Crist Collection (AFC/2001/001/84581), Veterans History Project, American Folklife Center, Library of Congress. https://www.loc.gov/item/afc2001001.84581/ 25:50 – 26:00.

dier in the nose of the plane would push the release button again. The two had to be perfectly in synch: Dale and the bombardier used an extended intercom cord for direct communication, and no one else could speak. The moment Dale positioned the blade of the screwdriver behind the bracket, he yelled "READY…NOW!" as he pushed on the screwdriver handle and the bombardier hit the button at the same time. The bomb released and descended safely out of the aircraft. "All you're doing is your job," Dale said.[204]

Whenever a B-29 was low on fuel or damaged after a mission, they would land at an airbase in Japan. While the ground crews worked on the plane, the combat crew were given a special pass allowing them to go into the Japanese towns in their flight clothes. The local people were friendly as the crew members visited bars and geisha houses. Crist says he hardly drank at all and stayed out of the geisha houses. "I was pretty good at playing poker and staying sober," he said. "I always had a lot of spending money."[205]

His last eight to ten missions were night strikes on "those d*** (Yalu River) bridges," he said, in which his aircraft was met with extensive antiaircraft fire. "You could see puffs of smoke, and you got hit every time," he said. "You had six or seven hours in the plane knowing, hoping, that tonight is not the night."[206]

During Dale's Korean War service, 20 missions earned you an Air Medal, and after 40 missions you could go home. But after his 40th mission, then Staff Sergeant Crist volunteered to fly with a second

---

204. Dale H. Crist Collection (AFC/2001/001/84581), Veterans History Project, American Folklife Center, Library of Congress. https://www.loc.gov/item/afc2001001.84581/ 28:00.
205. Dale H. Crist Collection (AFC/2001/001/84581), Veterans History Project, American Folklife Center, Library of Congress. https://www.loc.gov/item/afc2001001.84581/ 25:50 – 26:00.
206. Dale H. Crist Collection (AFC/2001/001/84581), Veterans History Project, American Folklife Center, Library of Congress. https://www.loc.gov/item/afc2001001.84581/ 50:20 – 53:05.

crew that had lost its tail gunner. He changed from his side gunner position to become a tail gunner under Captain Williamson as aircraft commander. That is when he got hit: "That shell hit right next to my left window," Dale said. "They kept telling us it was bulletproof glass, but that shell was so close it shredded the rudder and made two big holes on the other side of my compartment and bounced back into me and I got wounded."[207]

That, he said, is where his flashbacks come from.

At the time you try not to think about that. There again, you're just trying to do what you're trained to do. You become a fatalist. When you first go into combat, you ask yourself if there's a bullet out there that's got your name on it. Then you start worrying about that, and then the more you're in combat, the more you get shot at, the more you realize there d*** sure is a bullet out there that's got your name on it, so what do you do? Well, you say it could happen to you so you surround yourself with the best surroundings you can. You make sure you're on the best d*** crew you can that knows what they're doing. "You do everything you can to protect yourself, so I found a piece of sheet steel, and I put it under my butt on my little seat there in the tail compartment, so I didn't get shot from below, you know.[208]

Those thoughts went away after the war. "You didn't talk about it to start with, then when you get back to regular life, you didn't think about it during the day," he said. But at night, your nightmares come around, and you can't do a d*** thing about that either, and you start

---

207. Dale H. Crist Collection (AFC/2001/001/84581), Veterans History Project, American Folklife Center, Library of Congress. https://www.loc.gov/item/afc2001001.84581/ 50:20 – 53:05.
208. Dale H. Crist Collection (AFC/2001/001/84581), Veterans History Project, American Folklife Center, Library of Congress. https://www.loc.gov/item/afc2001001.84581/ 53:05 -55:25.

thinking to yourself, is it just me? Why can't I get over this? Later you find out that it's not you, but there's nothing you can do about it. then at night… then you tell yourself, well I'm gonna get over this and it does get better. …back in the old days, we called 'em flashbacks. Now they call it PTSD."[209]

Dale wanted to continue flying missions in Korea, but he says that General LeMay himself ordered him shipped back home to work with a Strategic Air Command Wing being formed for the atomic bombing mission. In 1953, SAC shipped him to the newly formed 308th Bombardment Wing at Hunter Air Force Base, near Savannah, Georgia. The Air Force promoted him to temporary Technical Sergeant and assigned him to be Wing Non-Commissioned Officer in Charge of gunners. He was discharged at the end of his four-year enlistment in October 1953.

He found a job in Sacramento, California delivering food to stores, then became a route man for what would become Frito-Lay in Sacramento, California. He grew that business into a major regional distributorship. In 1972, he seized an opportunity to buy a wholesale bar supply distributorship and again found success, retiring to his home in Colorado.[210]

---

209. Dale H. Crist Collection (AFC/2001/001/84581), Veterans History Project, American Folklife Center, Library of Congress. https://www.loc.gov/item/afc2001001.84581/ 55:25 – 1:05:40.

210. Dale H. Crist Collection (AFC/2001/001/84581), Veterans History Project, American Folklife Center, Library of Congress. https://www.loc.gov/item/afc2001001.84581/ 1:24 – 1:35.

Squadron Commander Awards Sgt Dale Crist with the Air Medal. Photo: Dale Crist. Source: Dale H. Crist Collection (AFC/2001/001/84581), (Veterans History Project, American Folklife Center, Library of Congress. https://www.loc.gov/item/afc2001001.84581/ 1:47:45)

Wing Commander presents Sgt Crist with the Purple Heart and his second Air Medal. Photo: Dale Crist. Source: Dale H. Crist Collection (AFC/2001/001/84581), (Veterans History Project, American Folklife Center, Library of Congress. https://www.loc.gov/item/afc2001001.84581/)

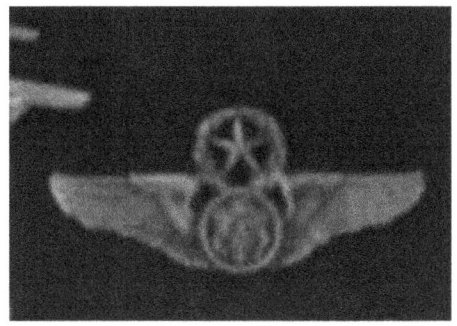

Chief Enlisted Combat Crew Badge worn by Staff Sergeant Dale H. Crist. Photo: Dale Crist, Dale H. Crist Collection (AFC/2001/001/84581), (Veterans History Project, American Folklife Center, Library of Congress. https://www.loc.gov/item/afc2001001.84581/ 1:50:48)

CHAPTER FIVE

Thomas W. Stevens

"I'm really glad I'm not down there."

Thomas W. Stevens' mother was in New York City visiting her mother when Tom was born there in 1933. His grandmother was in charge of housekeeping at a downtown hotel. His family, though, had a farm in Minnesota, and Tom and his mother soon returned there where he spent his early childhood. By the time he was in grade school, the family moved from Minnesota to Beloit, Wisconsin, and later to a farm there, where hard work and grown-up responsibilities dominated young Thomas' life.

The Stevens family eventually moved to another small farm near Ozark, Missouri, where the six-foot-one Tom played on the 1950 Ozark, Missouri, Class 2 high school basketball championship team.[211] Looking for a direction in life after graduation, he enlisted in the United States Air Force on July 12, 1951.

The United States was then ramping up military manpower to meet the war's need by restarting the draft and recalling reservists to active duty. When Tom reported for training in July at Lackland Air Force Base just outside San Antonio, Texas, he was initially housed in a large group tent. At the time, there were more recruits than barracks, so the Air Force erected a temporary tent city.

---

211. Missouri State High School Activities Association, "Boys Basketball State Championships," https://www.mshsaa.org/Activities/StateChampionships.aspx?alg=5

In his oral history, Tom tells of young Airman Stevens' first Air Force celebration event. After his flight of some 75 recruits moved into barracks and several long days of close order drill in the hot south Texas sun, the drill sergeant told them, "OK gentlemen, tonight we're going to have a party."

"I thought, man, that is very nice, the Air Force is gonna have a party for us! How wonderful," Tom said. The drill sergeant clarified that it would be a "G.I. party." Tom didn't know what a G.I. party was, but he learned that evening. "We were down on our hands and knees with brushes and soap and water scrubbing the floors of the barracks. That was the party."

Part of the drill sergeant's instruction was teaching the recruits how to make a bed with a hospital corner. "To this day I still put a hospital corner on the bed at home," Tom said. "I just don't feel right unless I do that. Why is that? I was only in the Air Force for four years. There are some things you pick up, and it just leaves an indelible impression on you. It affects the way you think and act the rest of your life."[212]

The Air Force tested all recruits during Basic Training to determine their aptitude for particular job specialties. Tom, however, concluded that what really determined placement was what the Air Force needed most at the time. So, after graduation from Basic, Airman Stevens was sent to Lowry Air Force Base near Denver, Colorado, for B-29 turret systems school.

He found Denver's climate more suitable to his Missouri upbringing than south Texas. The school was at times challenging, but Tom was more than up for it, meeting or exceeding all academic requirements in the four-month program. The aerial portion of gunnery training began about a month into that instruction. Tom said it was

---

212. Thomas W. Stevens, Korean War Legacy Project, 2024. https://koreanwarlegacy.org/interviews/thomas-w-tom-stevens/

exciting and not scary at all for him as an 18-year-old. Not the least of the rewards was flight pay, adding $37.50 to his $80.00 monthly basic pay.[213] Eligibility for flight pay required a minimum of four hours in the air each month. Tom had no problem passing through that gate.

He still was one last training step away from fighting in the Korean War. The Air Force sent him from Lowry Air Force Base back to San Antonio, this time to Randolph Air Force Base to join a newly formed B-29 crew. He was the final arrival, thus assigned to the least desired position: tail gunner. The aircraft commander was Captain Dinny J. Fotinakes, of Geneseo, Illinois.

Captain Fotinakes, a World War II veteran of 30 missions[214], had been recalled to duty for the Korean War, as had the crew's other four officers. In World War II, Captain Fotinakes began as co-pilot for one of several B-17 Flying Fortress bombers that would be nicknamed "Ice Cold Katie,"[215] and became pilot of one of many Flying Fortresses known as "Feather Merchant."[216] His 30 combat missions for the 709th Bombardment Squadron, 447th Bombardment Group, included the preparatory bombing prior to the invasion of Europe and D-Day strikes on the beachhead and nearby airfields. He came to the Korean War as a seasoned veteran airman, awarded the Distinguished Flying Cross and the Air Medal with three Oak Leaf Clusters in World War II[217]. Having also earned aerial gunners' wings himself

---

213. Military Pay had not changed since 1949. Service members would not see a raise until May 1952. Department of Defense, Defense Finance and Accounting Service, "Monthly Basic Pay and Allowances," effective 1 October 1949. https://www.dfas.mil/Portals/98/MilPayTable1949.pdf .
214. *The Dispatch*, Moline, IL, September 18, 1944, 14.
215. *The Rock Island Argus*, Rock Island, IL, March 17, 1944, 15.
216. "Gets Second Cluster," *The Rock Island Argus*, April 29, 1944, 18.
217. "Moline Pair Gets Flying Cross for Action in Europe," *The Rock Island Argus*, Rock Island, IL, July 11, 1944, 10.

during his air cadet training in the U.S. Army Air Forces, Captain Fotinakes also had high expectations for all his gunners.[218]

The crew's enlisted men were comparatively new to the Air Force, except Sergeant Leo Sommers, the flight engineer, who also had served during World War II. The crew worked together for several weeks on training missions from Randolph Field, Texas, to various locations around the U.S., but mostly within Texas. One mission took the crew to a simulated bomb run on a target in Springfield, Missouri, a flight that went over the family farm.

The full crew transferred to Forbes Field, Topeka, Kansas, for training flights involving heavy-weighted takeoffs. For this purpose, they loaded the B-29 with tanks filled with water instead of bombs. The added weight created the agonizingly slow and low takeoff that typified a B-29 combat mission with a full fuel and bomb load. When they reached Kansas farmland outside Topeka, they would empty the water tanks, providing free irrigation for some of America's heartland wheat fields. The farmers never complained about aircraft buzzing overhead.

The Fotinakes crew departed Forbes Field in mid-October for the long journey to Kadena Air Base on Okinawa off the southern coast of Japan. They traveled via Air Force Military Air Transport Service passenger aircraft. Enroute, they were separated onto differing transport aircraft, and Tom's group were able to spend a few extra days in Honolulu, Hawaii before moving on to Okinawa. It was quite an experience for a small-town Missouri farm boy. Airman Second Class Thomas Stevens and the others in the Fotinakes crew reported for duty to the 307th Bombardment Wing, 371st Squadron at Kadena Air Base on October 29, 1952.

The 371st had just moved out of tent city and was quartered in

---

218. "East Moline Air Cadet Will Be Awarded Wings," *The Dispatch*, Moline, IL, June 14, 1943, p. 7.

a long single story cement block building with double rows of cots. A few months later, they moved into newly constructed two-story buildings, two to a room and a common bath down the hall. Tom's roommate was crewmate and fellow Missourian Jerry Lewis, the right blister gunner.

Combat orientation in the 307th was rigorous in late 1952. Training and checkout were scheduled to take about three weeks. Most crews took longer. Gunners fired an average of 450 rounds of caliber .50 ammunition—more than they had previously fired during their entire stateside training courses. The full combat crew had to pass a final check ride under the scrutiny of the 307th's Standardization Evaluation checkout crews.

The most difficult crew task was to become proficient in radar bombing, a skill set that also involved gunners. The essential navigation and flying skills required the radar operator, navigator, bombardier, flight engineer, radio operator, pilot, and aircraft commander to manage a lengthy, complex challenge, keeping the aircraft following a narrow beam that arced from the ground and into the atmosphere, reaching across Korea. The crew kept the B-29 on one electronic beam until it intersected another from the other side of the peninsula. These beams were calibrated to intersect precisely over the target.

Gunners were responsible for visually monitoring the engines and control surfaces, confirming the bomb bay doors were open, and notifying the crew when U.S. escort fighters arrived, all the while keeping a sharp eye out for enemy fighters. The tail gunner had the best view of the exploding bombs below and provided an immediate bomb damage assessment.

The Fotinakes crew was the only replacement combat crew to arrive at the 307th in October. They were cleared for combat in less than 21 days.

On a combat mission, about an hour before reaching the combat zone, Tom would begin his preparations to move back to his solitary tail gunner compartment, donning his cold-weather clothes, Mae West inflatable life preserver, and parachute harness. He left one chest pack parachute close to the rear entrance door in the unpressurized area. If any tail gunner tried to bail out of the aircraft through the small escape hatch in the tail while wearing heavy clothing and parachute, he simply would not fit.

Once at his position, Tom made himself as comfortable as he could in a confined area with bulky attire. If the aircraft commander ordered the crew to bail out, Tom would have to exit the tail compartment, make his way around the lower aft turret, snap on his chest pack parachute and exit through the rear fuselage door. Tail gunners realized their chances of successful bailout from a crashing aircraft were not good. Tom and his fellow tail gunners did not dwell on that but always prepared thoroughly for any possibility on every mission.

In his oral history, Tom recalled that the 307th had suffered significant losses in a large aerial battle with MiG-15s a year before he arrived. Five of the 307th's bombers were destroyed in an encounter with more than 50 Soviet-piloted MiGs. But by the time the Fotinakes crew went into combat a year later, the B-29 community had developed countermeasures, both technical and doctrinal, that minimized the risks of deadly encounters with MiG-15s. "The MiG-15 against the B-29 was just like David and Goliath," Tom said.[219]

After the B-29 losses of April and October 1951, Bomber Command held a commanders' conference and decided to immediately implement innovative bombing techniques that were then being developed. Using secret radar equipment called SHORAN for Short Range Aerial Navigation, designed for navigating Strategic Air Command atomic bombers to targets inside the Soviet Union, the B-29

---

219. Thomas W. Stevens, Oral History, Korean War Legacy Foundation, https://koreanwarlegacy.org/interviews/thomas-w-tom-stevens/#clip-1

wings in the Korean War were able to navigate precisely to targets in North Korea at night.

Bomber Command also changed their tactics. Flying at night eliminated the need for close defensive formations. Instead, they flew the bomb run in single file, with about one minute between bombers. These technical and tactical innovations required the addition of a crewman to operate the SHORAN equipment. It also meant that coordination among the SHORAN operator, pilot, bombardier, and, especially, the navigator had to be immaculately precise. These innovations immediately reduced MiG-15 encounters almost to none.

Tom Stevens' first combat mission was memorable, historic, really. It was November 28, 1952, a nighttime maximum-effort mission by all three B-29 wings to strike targets along the Yalu River at Sinuiju and Uiju. The 307th's mission Number 470 involved 18 of the 307th's B-29's—all that were combat ready and had combat qualified crews available. They were scheduled to simultaneously attack six targets in the Yalu River corridor. Aimpoints included the Sinuiju Airfield, Sinuiju Locomotive Works, Uiju Airfield, Uiju Communications Center, and two Anti-Aircraft Batteries.[220]

As the 307th began its bomb run, Tom, with his panoramic view from the tail gunner position, could see the 20-some searchlights beaming into the night over the target area. But the 307th's electronic countermeasures operators totally jammed enemy radar, making the lights dance aimlessly in the sky. The enemy fired its flak, but only one B-29 took a hit, and that was minor.

"At any rate it was kind of dangerous because there were a lot of MiG aircraft stationed right on the Yalu that could have come up after us," Tom said. "But the MiGs were not a problem, it was primarily flak because of the night missions being flown." And while they

---

220. 307th Bombardment Wing History, October 1950, AFHRA Reel N0265, 832-833.

encountered a lot of flak, he said, "on the bomb run you can't take any evasive action, your main concentration is on hitting the target on the bomb run. As soon as the bombs were released, we'd make a steep descent and turn a 180 turn."[221]

The Wing's official history confirmed that most missions in this timeframe saw lots of flak but no MiGs. Tom's perspective was also consistent with Bomber Command's report: "As a matter of fact, the lack of effective enemy night fighter opposition had Bomber Command staff considerably worried as it was difficult to understand… The command during October (1952) was operating with almost complete operational freedom in good weather and in bad everywhere except of the Yalu."[222]

A similar report is in the November 1952 record. "We always had flak," Tom said. "It got pretty close. It didn't hit us, but it seemed like it was close. I was primarily interested in getting back alive."[223] He never wore the issued flak suit, but he did line the floor of the tail compartment with one to provide added protection for his posterior.

On the bomb run on that November 28, 1952 mission, something potentially catastrophic happened: the B-29's crossed the Yalu River into Chinese airspace over an area adjacent to the "Manchurian Sanctuary."[224] President Harry Truman had issued strict guidance that American aircraft were not to cross the Yalu River into Chinese airspace. American fighter jets were not even allowed "hot pursuit" of an attacking MiG for fear that doing so would prompt the Soviets to escalate the war perhaps to the point of using their own atomic bombs.

---

221. Thomas W. Stevens Collection (AFC/2001/001/86198), Veterans History Project, American Folklife Center, Library of Congress https://www.loc.gov/item/afc2001001.86198/
222. Bomber Command History, AFHRA Reel K7179, 417.
223. Thomas W. Stevens Collection (AFC/2001/001/86198), Veterans History Project, American Folklife Center, Library of Congress https://www.loc.gov/item/afc2001001.86198/
224. 307th Unit History, AFHRA Reel N0265, 832.

> "Technically we were not supposed to go into Manchurian airspace," Tom recalled. "I do recall once when we were turning having dropped the bombs, there was a strong wind, and it blew us into Manchurian airspace. That was kind of—not a joke—but something that the officers laughed about because we were not supposed to be over there, we better get back over the Yalu River."[225]

After dropping bombs on target, if the flight engineer determined on the return flight that the aircraft was low on fuel or had a mechanical problem, the aircraft commander could decide to land at Itazuke Air Base in Japan to take on fuel before heading back to Okinawa. The crew welcomed the short diversion because they could have a meal at the dining facility there and enjoy fresh eggs.

When they finally arrived at Kadena after each mission, the crew would be picked up by a truck and taken to an area for debriefing. Everyone was offered a cup of whiskey. The 307th wing commander offered a bottle of whiskey to crews placing bombs in a perfect scoring pattern. "I'm not sure how accurate some of the information was after a few drinks of that whiskey that they gave us," Tom Stevens said. "As I recall it was called Old Methuselah; I don't know if that's still on the market or what or where they got that, but I do recall pouring some into a paper cup and picking up the cup and the bottom fell out, so it was pretty strong stuff."[226]

The next day the crew would be back on the flight line cleaning and assisting the ground crew with maintenance. Machine gun maintenance in the intense humidity and salty air of Okinawa was tedious

---

225. Thomas W. Stevens Collection (AFC/2001/001/86198), Veterans History Project, American Folklife Center, Library of Congress https://www.loc.gov/item/afc2001001.86198/
226. Thomas W. Stevens Collection (AFC/2001/001/86198), Veterans History Project, American Folklife Center, Library of Congress https://www.loc.gov/item/afc2001001.86198/

and time consuming. Keeping the machine guns free of rust was a constant battle. Rapid corrosion onset did a number on the guns and turrets. To make matters worse for gunners, the wing had a critical shortage of M3 machine gun replacement parts and headspace gauges,[227] and the only way to solve that would be assigning the wing's gunnery requisitions higher priority, a decision that only higher headquarters could make.

Tail Gunner Staff Sergeant Thomas W. Stevens completed 27 combat missions and returned to the United Sates in Spring 1953. "I think we did pretty good," he said.[228] For all the risk in fighting in the air, Tom knew that the soldiers on the ground faced hardship and danger surpassing the exposure of B-29 combat crews. "When we flew over Korea, we could see the mortars and artillery shells being fired on the ground. I thought to myself, I'm really glad I'm not down there. I had the utmost respect for the guys fighting on the ground."[229]

He stayed in the Air Force and retrained as a B-50 waist gunner stationed at Biggs Air Base, El Paso, Texas. Strategic Air Command then transferred him to Walker Air Force Base near Roswell, New Mexico, as a gunner for the nation's atomic bomber force until 1955. By then the B-29 and its successor B-50 were replaced by the all-jet-engine B-47, followed shortly by the B-52. Fewer gunners were needed for bombers, so Tom chose to accept an honorable discharge when he completed his four-year enlistment.

In 1955, he enrolled in Springfield Missouri's Drury College, where he met and married Barbara. On graduation, Southwestern

---

227. These gauges are pieces of tool steel, cut to precise lengths. They're shaped like cartridges, so they fit in the firearm's chamber. The firearm's bolt should close normally on the "go" gauge and should not close on the "no-go" gauge.
228. Thomas W. Stevens Collection (AFC/2001/001/86198), Veterans History Project, American Folklife Center, Library of Congress https://www.loc.gov/item/afc2001001.86198/
229. Thomas W. Stevens, Oral History, Korean War Legacy Foundation, https://koreanwarlegacy.org/interviews/thomas-w-tom-stevens/#clip-1

Bell Telephone Company hired him into their management program in St Louis, Missouri, from which he retired after 33 years as district manager. Tom and Barabara raised a family of two sets of twins (girl twins followed four years later by boy twins), resulting in seven grandchildren and three great-grandchildren.[230]

But Tom was not finished with his service to the nation. In 2003, he and all the members of the Kansas City Korean War Veterans of America in Johnson County, Kansas, championed efforts to build a Korean War Veterans Memorial in Overland Park. The memorial was dedicated on September 30, 2006, in a ceremony attended by over 1,000 people.

In 2010, he was elected to a three-year term on the National Board of the Korean War Veterans of America and served a second term for a total of six years. When the Air Force reactivated the 307th Bombardment Wing, the Air Force invited him and several other 307th "Old Timers" to participate in the reactivation ceremony at Barksdale Air Force Base, near Shreveport, Louisiana, on January 8, 2011.

In 2016, Tom was voted in as national president of the Korean War Veterans of America. In that position he championed the cause of Korean War veterans in meetings with President Barack Obama, Vice President Mike Pence, Secretary of Defense James Mattis, and South Korean President Moon Jae-in. "South Korean officials in the U.S. and South Korea have always treated Korean War Veterans with honor and dignity," he said. His term as president ended in June 2018, but Tom continues to work with the Korean War Legacy Foundation and other Korean War veterans' groups to advance Korean War legacy and education programs in American schools. In 2018, Drury University recognized Tom's lifetime of service by awarding him "Distinguished Alumni, Community Service."

---

230. Thomas W. Stevens, Oral History, Korean War Legacy Foundation, 2017. https://koreanwarlegacy.org/interviews/thomas-w-stevens-2nd-interview/

The result of the Korean War, he said, "was that South Korea is a democracy. The efforts on the part of the United Nations forces resulted in South Korea being free today."[231] Tom notes that the Korean people continue to express their sincere appreciation for what the United States did to keep their home country free from the evil of communism, and he is sometimes amazed at the extent to which they go to show their appreciation. He urges all Americans to educate themselves on the legacy of the Korean War and never take freedom for granted. For as the inscription on the Overland Park, Kansas Korean War Memorial declares:

"Freedom is Not Free!"

---

231. Callie Counsellor, "Kansas City-area Korean War Veteran Reflects on Service During 4th of July," KSHB 41 News, July 5, 2021. https://www.kshb.com/news/local-news/kansas-city-area-korean-war-veteran-reflects-on-service-during-4th-of-july

The Fotinakes crew at Randolph Field, San Antonio, TX before departing for Okinawa. Captain Fotinakes is kneeling in the front row, third from the left. Tom Stevens is in the back row, first on the left. Photo: USAF, courtesy Thomas W. Stevens

\

Members of the Fotinakes crew at Kadena Air Base. Left to right: Leo Sommers, Flight engineer; Dinny Fotinakes, Aircraft Commander; Jerry Lewis, Left Waist Gunner. (Photo courtesy Thomas W. Stevens)

Stevens on the flightline at Kadena Air Base. Gunners shared duties for checking bombs before they were loaded. Once the aircraft was airborne, gunners also climbed into the bomb bays and armed the bombs. (Photo courtesy Thomas W. Stevens)

Stevens with a bucket of solvent cleaning rust and salt off his tail guns. The flash hiders at the end of each barrel were added to reduce the chance that an enemy fighter could zero in on the B-29 at night. (Photo courtesy Thomas W. Stevens)

Lt Col Ernest Turner, Commander 371st Bombardment Squadron, presents Sergeant Thoms W. Stevens with the Air Medal for completion of 27 combat missions in the Korean War. (Photo: USAF courtesy Thomas W. Stevens)

S/Sgt Stevens in the right waist gunner's seat of a B-50 Bomber. At the end of the Korean War, Stevens was assigned to Walker AFB, Roswell, NM where he transitioned to the B-50 bomber, a B-29 with more powerful engines and several modifications to enable the aircraft to deliver atomic bombs on the Soviet Union during the Cold War. He also flew on B-50s from Biggs Air Force Base in El Paso, Texas before his enlistment was up in 1955. (Photo courtesy Thomas W. Stevens)

The Korean War Veterans Memorial, Overland Park, Kansas. The Memorial was created with efforts of the Korean War Veterans of America Chapter #181 and features the bronze statue "Remembering," by Charles Goslin. Eight of the thirty granite panels feature the names of 415 Kansans who were lost in the war. (Photo: "Korean War Veterans Memorial," Overland Park, Kansas https://www.opkansas.org)

KWVA National President Thomas W. Stevens and wife Barabara with President Obama at the White House, Veterans Day 2017. (Photo: Courtesy Thomas W. Stevens)

KWVA National President Thomas W. Stevens meets with Vice President Mike Pence and Defense Secretary James Mattis prior to Stevens' address at the 2017 Veterans Day Ceremony at Arlington National Cemetery. (Photo Courtesy Thomas W. Stevens)

CHAPTER SIX

Romaine Gregg

"It's really cold and dark in the tail."[232]

Only 16 years old, Romaine Gregg enlisted in the Air Force in August 1951. It was a patriotic act on his part when the Korean War broke out, but his primary motivation was to get away from Peotone, Illinois, a small farming village 40 miles south of Chicago. In 1951, everyone knew everyone else in Peotone, and they all knew about the kid being raised by his grandma.

Romaine already was a master mechanic. He could build or fix just about anything. He had helped his Uncle Guy build Grandma's home, learning how to cut steel, weld, and pour concrete. He mastered the crafts of framing, siding, roofing, electrical wiring, and plumbing. All 1,200 residents of Peotone saw the wiry kid muscle, strain, lift, shape, and move God's raw resources and man's fabricated materials into a house of strength and pride.

Romaine had one weakness, which everyone in Peotone knew, too. He could barely read. He squeaked through his sophomore year at Peotone High School and knew he would flunk out of school in his junior year. Reading was more painful for Romaine than pulling teeth for this otherwise fearless 16-year-old—fearless except, perhaps,

---

232. This chapter relies in part on interviews conducted by the author with Romaine Gregg in the summer of 2024.

for the prospect of repeating a year of school. But he also knew that Grandma would not let him drop out.

So, while he spent his post-sophomore-year summer working in his Uncle Rick's steel fabrication shop, he checked in at the Armed Forces Recruiting Office. The Korean War was on. He heard the war stories of guys with less than an eighth-grade education who fought in World War II. Maybe he could put his skills to work in the military while getting himself out of Peotone.

He was one year too young to enlist with parental permission. But the Air Force recruiter was willing and able to alter his birth certificate. Now Romaine had to beg his mother and father to sign the release required for anyone under the age of 18 to enlist. They relented.

One quiet morning in August 1951, the recruiter picked Romaine up at Grandma's house at 5 a.m. They drove off before anyone knew he was gone. Romaine was grateful for all that his grandma and uncles had done for him. He'd call them later from the Air Force. He was sure they'd be proud of him.

He and the recruiter rode in silence for the hour-and-a-half drive to downtown Chicago. There, recruits hustled through several stations in a building for a medical exam and shots. Next was a bus trip to a YMCA where Romaine would spend the night in a third-floor room on the north side of the building. He remembers the area below as filled with "burlesque shows, alcoholics, and fights on the sidewalks." Strange scenes for a small-town farm boy.

The next morning, reception station sergeants woke the recruits at 6 a.m. for breakfast. Then they filled out questionnaires. Romaine struggled to read the words but got through them with help. After an interview, Romaine took the oath in the afternoon before boarding a bus to the train station for the long cross-country journey to Lackland Air Force Base on the western side of San Antonio, Texas. The first station at the initial reception center was for uniform issue. By the time Romaine got to the head of the line, the only available trousers were standard issue 34-inch waist. That was too large for his wiry shape, so he had to report to an array of tailors who took in his trousers to fit his 28-inch girth.

At the end of World War II, the Army Air Forces had more than two million airmen.[233] By June 1950, what had become the U.S. Air Force as a separate service had about 400,000.[234] The Air Force consolidated its Basic Military Training at Lackland. But when the Korean War broke out, the Air Force had to expand rapidly and opened a second facility at Sheppard Air Force Base in Wichita Falls, Texas. That afternoon, Romaine and the other recruits with him travelled by train north from San Antonio to Wichita Falls.

The next morning the men enjoyed a Sheppard Air Force Base "special" breakfast of powdered scrambled eggs, two pieces of cold toast, a cup of fruit, two sausage links, a cup of coffee, a small glass of juice, and a larger glass of milk. For the next two months, every day was a "special" breakfast day. Romaine found the rigors of Basic Training to be tolerable, as most of it was physical and mental strengthening. There was not much reading.

He distinctly remembers his experience in the tear gas tent. The men were required to remove their gas mask and recite their name, rank, and serial number while gasping for breath in the wretched fumes before the sergeant allowed them to exit. Outside, most men were bent over, spitting up and gagging as tear ducts gushed. Romaine had held his breath the moment he removed his mask, exhaled while doing the recitation, and dashed out the doorway as the sergeant cleared him to move out. His brief exposure to tear gas seemed not

---

233. Headquarters, Army Air Forces, *Army Air Forces Statistical Digest, World War II* (Washington DC: Office of Statistical Control, December 1945), 15. https://media.defense.gov/2011/Mar/31/2001330134/-1/-1/0/AFD-110331-045.pdf

234. Headquarters, United States Air Force, *United States Air Force Statistical Digest Jan 1949 – Jun 1950* Fifth Edition (Washington, DC: Operations Statistics Division, D/Statistical Services, DCS/Comptroller, 25 April 1951) 28. https://media.defense.gov/2011/Apr/05/2001329940/-1/-1/0/AFD-110405-027.pdf

much more powerful than the fumes from an oxy-acetylene torch in the small steel cutting shed of his uncle's shop.

Upon graduating from Basic Training, the trainees could select a job preference. Romaine had flown in a Piper Cub just before enlisting in the Air Force, infecting him with the flying bug. He knew he could not be a pilot, but he told them he wanted to fly. His demonstrated mechanical skill made it an easy call for the Air Force: they sent him to gunner's school at Lowry Air Force Base in Denver, Colorado.

Romaine quicky mastered the B-29 gunner's expertise. Dismantling and reassembling the M3 machine guns, tracing the complex electrical wiring from the sight to the computer to the turrets, the complicated machinery for the turrets, even the rudimentary analog computer, all came naturally to him. They paralleled the skills he picked up through his work back home.

His main difficulty was reading a technical manual the night before a particular class. But his motivation to earn his gunner's wings and to fly drove him to work through it. He stayed up late at night in the latrine, tracing the text by finger and reading aloud word-by-word, to gain enough familiarity so that he could more quickly grasp the instructor's lesson in class. Classes usually started with a demonstration, often using a mock-up or illustration of the actual equipment. Romaine only needed to see it done once to do it himself. "It's not really a complex system, once you get into it and how it operates," he said.

He thrived on the additional duties of the gunner. Before the flight engineer and aircraft commander could start the aircraft's engines, the tail gunner had to start up the auxiliary power motor, the "putt-putt." Even when the main engines cranked up, they had to be turning at least 2200 rpm before they could start generating full power. The putt-putt drove the B-29s braking and hydraulic systems until the engines were fully revved up. Romaine knew well how a small engine could quickly consume fuel, and he never failed to ensure that a 2-gallon can was filled with spare fuel, just in case. When the main engines cranked up, he was required to be on the lookout for any signs of leaks, smoke, or fires inside and out.

There were plenty of other pre-flight duties for the tail gunner. He checked the nose gear linkage and nose wheel nacelle doors. He inspected each main landing gear tire and brake shoes, plus the navigation and landing lights. On top of the wings, he was responsible for checking the fuel level in all four wing tanks, deicer fluid for the four propellers, and deicer boots on the leading edge of the wings as well as the turret dome locks. As he made his way to the tail of the aircraft, he checked the deicer boots on the vertical and horizontal stabilizers.

Afternoons were given to more physical duties, especially bailout procedure. After learning how to do the classic military parachute landing fall from three-foot platforms on the ground, they moved up to the 40-foot tower. This was the closest thing in training to an actual parachute jump. Airmen practiced leaping out into the air in a full harness suspended from straps connected to an overhead cable that took them along like a zip line to a spot closer to the ground where the jumper has to make a good landing.

Then they did it over water, wearing a yellow rubber vest inflated using CO_2 cartridges, with a back-up manual inflation tube. They also had a one-man inflatable life raft strapped to their butt. Romaine was an excellent swimmer and thrived on the thrills of jump school.

Romaine Gregg took his first flight on January 25, 1952. It gave him a tremendous rush. "I will never forget the exhaust smell of those R-3350's!" he said of the engines that powered the B-29 Superfortress. Romaine received his Gunner's Wings on February 17, 1952, his 17th birthday. The thrill was tempered by an accident he witnessed during flying week. A B-29 lost an engine in flight and crashed to the ground. The three men in the front compartment bailed out, but the three enlisted men in the aft did not.

From Colorado's Lowry Air Force Base, the Air Force transferred Romaine back to Texas, this time to Randolph Air Force Base on the eastern side of San Antonio. There, he was assigned to a newly forming combat crew. Combat crewmen became as close to each other as family. The enlisted men were like brothers: Tail Gunner Romaine Gregg, Radio Operator Glenn Haltom, Right Gunner Wayne Walker, Left Gunner Stan Baumgarner, Central Fire Control Gunner William Murphy, Radar Operator Bob Nicosia, and Flight Engineer Gene Galloway. They lived in the same barracks bay, ate together in the mess hall, and hung out together off-duty.

Gene Galloway was the senior and most experienced member of the crew. He took the young tail gunner under his wing. "My flight engineer was the father figure in my life," Romaine said. "He taught me the proper things a grownup does. Like never let a woman pay the check, open the door for a woman, don't let a door slam in another person's face. I grew up and became a man in the Air Force, all thanks to his guidance."

It took several months for Romaine's crew to deploy to the combat zone. There were never-ending delays getting some airmen across the Pacific Ocean because the United States was deploying so many troops on so few transport aircraft. Most enlisted men travelled by ship. Only one U.S. Navy transport departed each month out of Camp Stoneman, California, in the San Francisco Bay. That boat transported soldiers, sailors, and Marines as well as airmen to and from the combat one.

Early in the war, the Air Force instituted a rotation policy to help prevent the kind of combat fatigue and burn-out that had severely afflicted aerial combat crews in World War II. Today the condition is known as Post Traumatic Stress Disorder. Beginning in 1951, the Air Force policy was that combat crewmen would serve six months in combat then be replaced by a fresh crew from the United States, known as the "Zone of the Interior." But the shortage of transport stretched that time to as much as nine months or more. Sergeant Romaine Gregg did not arrive in the combat zone until December 1952.

Romaine was among nine replacement crews and 11 "filler" combat crewmen to report that month to the 371st Squadron, 307th Bombardment Wing on Okinawa. It was balmy, and the typhoon season had ended in November. After going through a three-and-a-half-day initial orientation, gunners went through a week-long refresher course on the remote-control turret system. He was then qualified to fly combat missions.

On one mission, his crew attacked the enemy airfield at Sinuiju. This was a sod runway that ran alongside the Yalu River in the heart of what was ruefully known as MiG Alley, only a few miles from the Soviet airfield at Antung, China. On this mission, someone fired rockets at the B-29. They were not the classic orange-glow "golf-ball" like artillery shells rising from the ground. Nor were they glowing red ballistic cannon tracer rounds from a MiG-15. These munitions trailed a horizontal yellow streak of flame from the tail-end of a rocket motor propelling a lethal explosive munition. From his panoramic perch in the tail, Romaine had a bird's eye view of the first and only time B-29 crewmen reported aerial rocket fire coming at them in a combat mission. What deadly new communist aircraft was launched at them; the intelligence officers would have to figure out.

Many of Romaine's missions took the crew to bridges on key choke points in the labyrinthine rail network that laced its way from the Yalu River bridges southward to Pyongyang and beyond to supply Chinese forces with food, fuel, ammunition, and replacement troops for the tens of thousands being captured or killed by United Nations forces along the line of contact. It was the "Spring Thaw," a time when warming temperatures melted snow and ice, raising the water to above flood level in rivers and streams, inundating the troop areas and lines of communication running north to south. There were about a dozen or so bridges tailor-made for the B-29 precision bombing techniques that Sergeant Gregg's 307th Bombardment Wing had mastered.

In fact, the 307th consistently topped the charts in bomb accuracy. In March, Romaine's B-29 was among 18 that knocked out all three principal bridges at Yongmi-Dong. A week later, they followed with strikes on the Sinanju railroad bridge amid intense flak. The flak was rendered ineffective by chaff and jamming, but Romaine's tail section was illuminated for two minutes guided by radar beams randomly scanning the skies. On this night, no MiG-15 Night Fighters emerged to follow the light beams to an air-to-air fight with the ever-ready wiry gunner from Peotone, Illinois. Though they hit their targets, B-29 bomb runs had little effect on the enemy's supply columns. Post-strike aerial photography revealed the bridges were most often repaired and back in operation the next day.

Bomber Command turned to precision daylight bombing of irrigation dams and hydroelectric plants to ruin the spring rice crops and flood the supply routes leading to Pyongyang. The first bombs were long fuzed to penetrate the 75-foot-high earthen dam and break it up. The following bombs were short fuzed to cause craters that would result in a breach under the pressure of the water behind the dam. The 307th struck the Kuwonga Dam with precision. But previous strikes by fighter-bombers alerted the dam operators to their vulnerability, and by the time of the B-29 strike they had lowered the water level, reducing the pressure on the earthen dam sufficient to prevent a breach.[235]

On June 14, Romaine's 307th departed Kadena Air Force Base on a mission to take down the Kusong Reservoir Dam to flood the airfield at Taechon. As the 307th zeroed in on the dam, the Chinese launched a massive surprise attack on South Korean positions along the 38th parallel at a location designated "Outpost Harry." Far Eastern Command diverted the 307th to support ground troops in this desperate battle.[236]

---

235. Bomber Command History, AFHRA, Reel K7181, 841.
236. 307th Bombardment Wing Unit History, June 1953, AFHRA Reel N0265, 1739.

B-29 Bombers are almost never diverted from high altitude precision night-time radar missions to low altitude, map-oriented, close-support bombing. Navigators had the difficult challenge of discarding all their pre-calculated waypoints and identifying new ones after digging out the correct aerial charts. They had no computers, no GPS signals, no radio beacons: just slide rules, pencils, brain power, celestial navigation, and dead reckoning. Then the flight engineers had even less time to break out their weight and balance tables and their "whiz wheels" to figure out whether the B-29 had sufficient fuel for the new mission and how to adjust the engines accordingly. Radio operators had to look up the frequencies of ground controllers, while pilots had to descend to lower altitudes, disengage the auto pilot, and fly the massive aircraft with seat-of-the-pants yoke and rudder controls.

The bombardier dropped bombs when the ground controller, using radar, told him precisely when to release them over enemy positions. Tail gunners like Romaine went along for the ride, but had an up-close and personal view of the effects the bombs had as the aircraft passed beyond the target.

Romaine was treated to quite a display of firepower on that mission. The communists were repulsed by the combination of the effects of the bombing and the heroism of the ground troops. Patrols sent into the area later reported thousands of enemy bodies amid smoking, smoldering ruins of equipment, supplies, and ammunition stocks.[237] A U.S. Marine was awarded the Medal of Honor, and a Greek infantry platoon became the only foreign unit to receive the U.S. Distinguished Unit Citation.

As the Armistice Talks reached a culminating point in June 1953, Bomber Command focused its strikes on the half-dozen enemy airfields just south of the Yalu River. Air Force intelligence had learned

---

237. Walter G. Hermes, *Truce Tent and Fighting Front* (Washington, DC: Center for Military History, 2012) 469.

that the Russians and Chinese were preparing to fly hundreds of MiG-15s to those airfields. When forces were frozen in place on signing the ceasefire instrument, the communists wanted a thousand or more advanced jet fighters in North Korea, ready to conduct a lightning strike to the south when the communists concluded the time was ripe to complete their takeover of the entire Korean peninsula.[238]

Romaine found himself among the bomber fleet attacking the infamous Namsi Airfield where, in October 1951, his 307th Bombardment Wing had lost five B-29s during a daylight strike in the most consequential air battle of the Korean War. On that fateful day while Romaine was in Basic Military Training, 150 MiG-15s swarmed over the 307th's nine aircraft flying in a tight formation in broad daylight. The U.S. had fewer than 40 jet fighters to protect the B-29s, and those American jets were either outmaneuvered or lured away from the main fight by some of the Soviet Union's top pilots, many of whom were aces from World War II.

This time, however, the 307th had developed advanced tactics, techniques, and procedures that resulted in complete mission accomplishment. On Namsi's 7,100-foot-long runway, Romaine and his fellow "Long Rangers" destroyed dozens of parked MiGs and the revetments where they were parked. Not a single B-29 suffered damage from flak or fighters.[239]

Sergeant Romaine Gregg's final mission was one month before the armistice took effect. On the night of June 26, 1953, the 307th destroyed the largest supply area in North Korea, located at Kuryong. The night attack was designated a "Cloak" mission. Post-strike

---

238. Robert F. Futrell, *The United States Air Force in Korea 1950-1953*, Revised Edition, (Washington, DC: Office of Air Force History United States Air Force, 1983), 679-684.
239. 307th Bombardment Wing Unit History, June 1953, AFHRA Reel N0265, 1739.

intelligence photos showed that nearly 250 buildings were destroyed. Personnel shelters and slit trenches were demolished.[240]

Romaine came home in 1953. He was just 19 years old with adrenaline still flowing in the full vigor of youth. He was discharged in December 1953 and returned to Peotone a changed man. He went back to work in general contracting but found an outlet for his energies in auto racing. He souped up a '39 Ford Coupe and won trophies in professional-level races at Chicago's Soldier Field[241] and the 87th Street Speedway.[242] He raced his '47 Ford at Soldier Field races, taking a 10-lap heat on June 25th, 1954, in a race that was won by future NASCAR Grand National racing giant Tom Pistone.[243]

Romaine earned his private pilot's license in 1972.[244] By 1960, he owned his own general contracting business and continued to work into his 80s, retiring in December 2022.

Romaine Gregg, the teenager who could barely read, is today a published author and illustrator, with three children's books in print. "I am a father of four and a grandfather of 10, and I have had the pleasure of putting many children to bed," he said. "In my early days, I told the children stories because I am a terrible reader. Recently, I thought it would be nice to put some of these fictional stories in book form so they could be enjoyed by others. That's how I became an author of children's books."[245]

---

240. 307th Bombardment Wing Unit History, June 1953, AFHRA Reel N0265, 1741.
241. "Pistone Wins Soldiers' Field Stock Feature," *Chicago Tribune*, June 26, 1955, 41.
242. "Duvall Wins Feature at 87th St.; Carey 2d," *Chicago Tribune,* September 1, 1954, 47.; "Crowe Scores in 87th Street Jalopy Event," *Chicago, Tribune,* May 31, 1955, 38.
243. "Pistone Wins Soldiers' Field Stock Feature," *Chicago Tribune*, June 26, 1955, 41.
244. Wayne Thomas, "Plane Talk," *Chicago Tribune* May 14, 1972, 73.
245. Romaine Gregg, *Kooter (A Frog)* (Pittsburg, PA: Rose Dog Books, August 2008). https://www.amazon.com/Kooter-Frog-Romaine-Gregg/

Sgt Romaine Gregg at his B-29 Tail Gunner Position. (Photo Courtesy Romaine Gregg)

---

dp/1434991172/ref=sr_1_1?crid=15FP0837UF2E&dib=eyJ2IjoiMSJ9.
tJSwsk6sKGjNXLmDqtKRAdsDf_E6GcccdvXZEUBvw1ldwB7A7jn-
hPAsIsTbXOUHayD3JuauYMI0rvmETmoH61QmmdVx6K9h_V_Fr-
5wLdEb28pw9X0ZYMvlwXltD1KOezGU1MZfevW0ap-K9Cf_XQRg.
QyHmazG_yvMm5sZGyIOUZjDE9-QTsEwDKeZvcvW9gk8&dib_tag=se&key-
words=romaine+gregg+books&qid=1727039432&sprefix=romaine+gregg+books%-
2Caps%2C218&sr=8-1

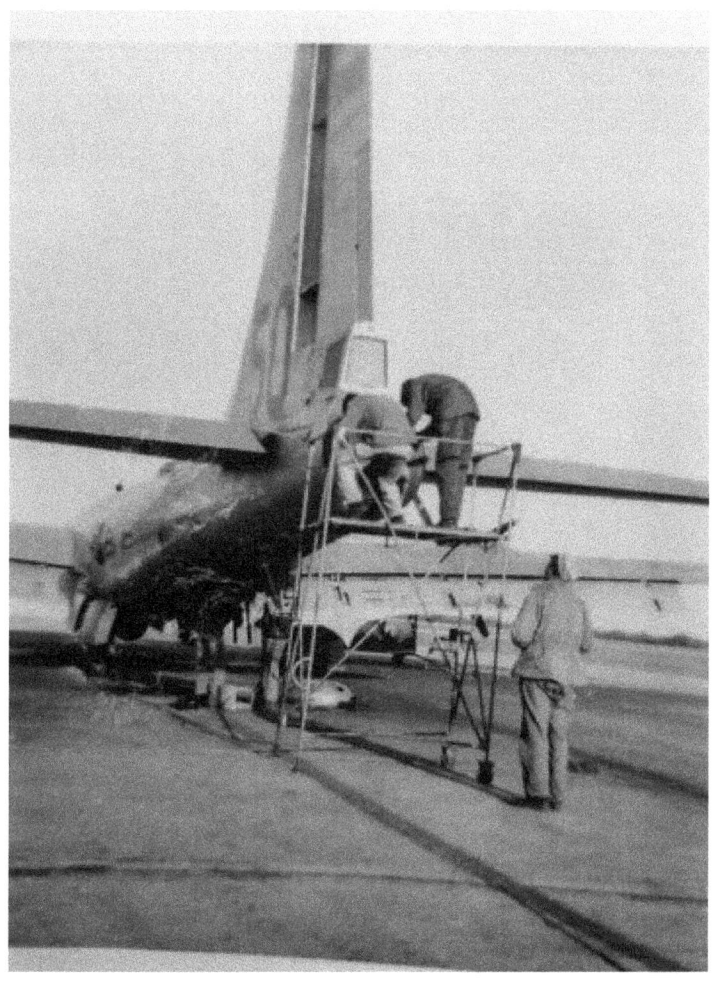

Sgt Gregg and another student servicing the turret of a B-29 under the watchful eye of an instructor in the Gunners' Course at Lowry AFB, Colorado. (Photo Courtesy Romaine Gregg)

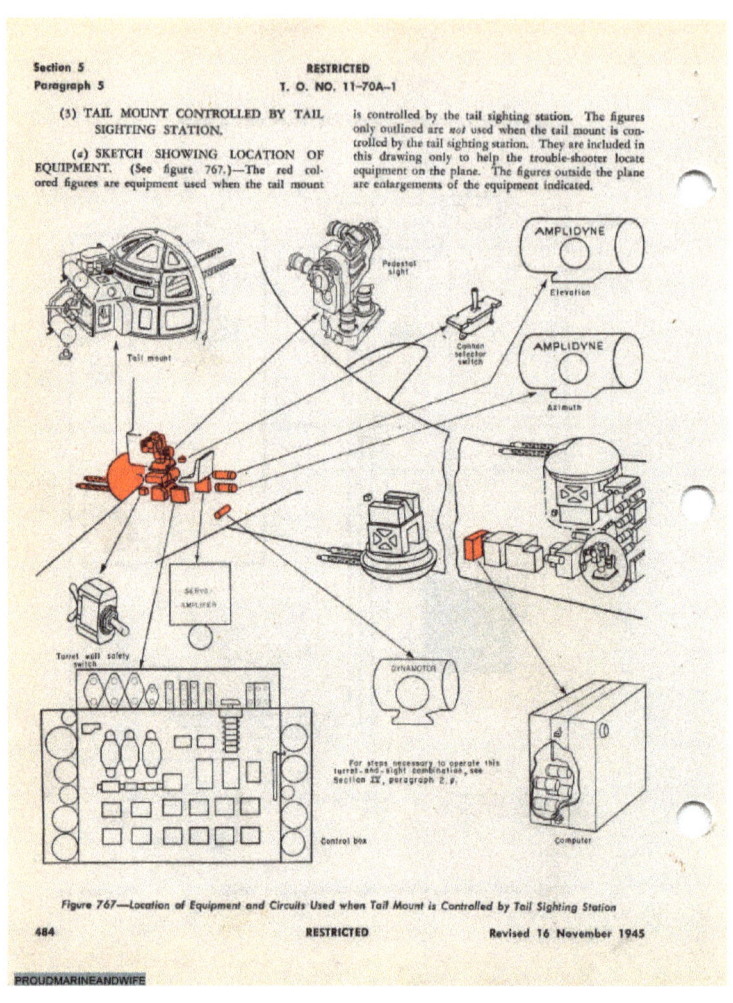

Diagram showing breakout of the complex components of the tail gun system that Romaine Gregg mastered. (United States Air Force, Handbook, Operation and Service Instructions: Central Station Fire Control System T.O. No. 11-70AA-9 (Washington, DC: Authority of the Secretary of the Air Force, 27 March 1951), 484)

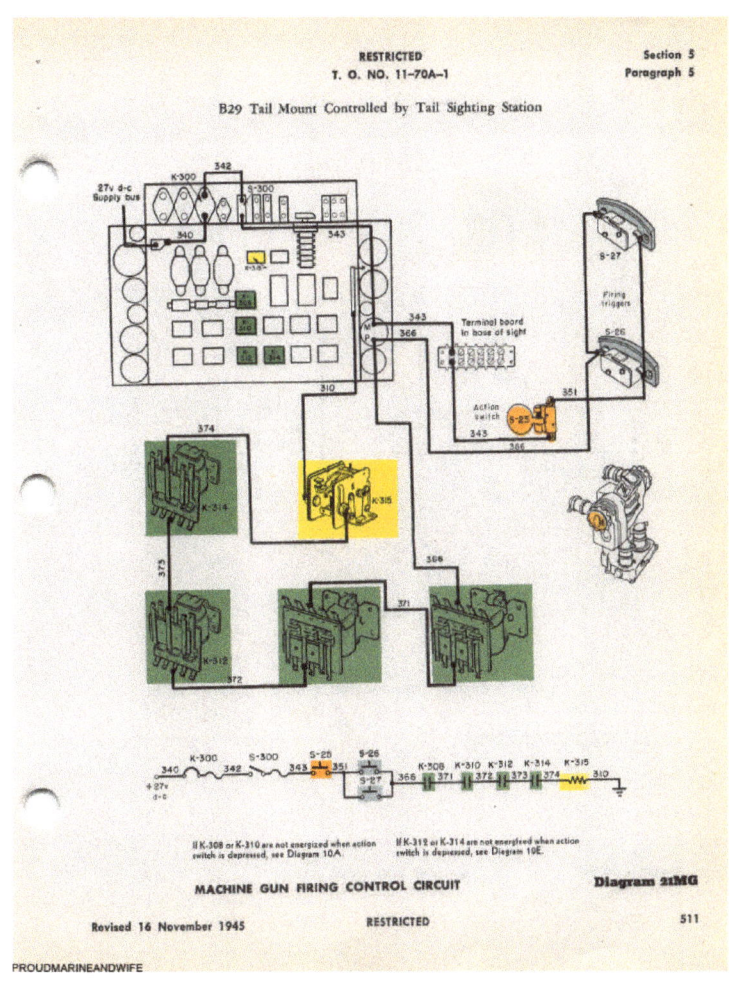

Wiring diagram for tail gun illustrating current flow from the gunner's trigger button to the bolt release that initiates the rapid firing of the M3 machine guns at the gunner's control. This is one of 32 circuits that Romaine Gregg, who could barely read, mastered in gunners' school. (United States Air Force, Handbook, Operation and Service Instructions: Central Station Fire Control System T.O. No. 11-70AA-9 (Washington, DC: Authority of the Secretary of the Air Force, 27 March 1951), 511)

Sgt Romaine Gregg relaxes between missions outside his barracks at Kadena AFB. (Photo courtesy Romaine Gregg)

Sgt Gregg inspects a bolt from his tail gun's M3 caliber .50 Machine Gun. (Photo courtesy Romaine Gregg)

Romaine Gregg Carries the checkered flag after taking the 10-lap heat race in his '37 Ford Coupe at the 87th Street Speedway August 31, 1954. (Photo courtesy Romaine Gregg)

Romaine Gregg shows off his rebuilt 1932 Ford Coupe. (Photo courtesy Romaine Gregg)

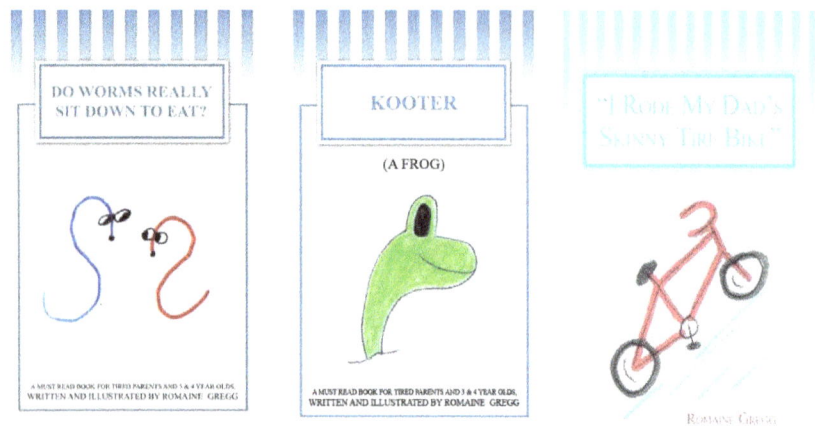

Three Children's Story Books Written by Romaine Gregg, 2008 – 2011

Well into his '80's, Sgt Romaine Gregg retains his crisp hand salute as he visits the World War II Memorial in Washington D.C. (Photo Courtesy Romaine Gregg)

CHAPTER SEVEN

Jack Bernaciak, The Last B-29 Combat Mission of the Korean War

Airman Second Class Jack Bernaciak took his seat in the 307th Bombardment Group mission briefing room at 0900 on July 25, 1953. It was Saturday morning at Kadena Air Base in Okinawa. That meant the mission would not end until late in the day on Sunday. What a way to spend a weekend in the tropics. Despite weeks and months of generals and politicians generating rumors about an approaching agreement to end the fighting, the 307th's combat crews had no time to relax.

Three months earlier, in April, the United Nations and communist forces had exchanged sick and wounded prisoners of war. But even as armistice negotiations moved slowly toward a cessation of hostilities, the communists stepped up their ground offensives in an attempt to seize more terrain farther south before the fighting froze front lines in place. Between April and July 1953, pitched battles produced 135,000 casualties on the communist side, 64,703 among United Nations forces.[246]

Air Force intelligence had also detected hundreds of MiG-15s

---

246. Walter G. Hermes, *Truce Tent and Fighting Front*, (Washington. DC: Center of Military History United States Army, 1992) 470-478.

appearing on enemy air bases south of the Yalu River in North Korea, clear evidence the communists wanted sufficient air power in place for a future air superiority campaign sometime after the armistice. In June, Far East Air Force Bomber Command's B-29s were put to work in a maximum effort to destroy as many of those airfields and jet fighters as they could.[247] By the end of June, the bombers had effectively neutralized all enemy airfields in North Korea.[248]

However, the communists had rehabilitated most of their airfields within a few days. They not only moved more fighter jets there, but they also parked many of them in earthen revetments that provided some protection from the 100-pound bombs that Bomber Command used. American B-29s changed their bomb load to 500 pounders to create more runway and bunker destruction than even the North Korean slave laborers could repair in short order.

By the third week of July, North Korean airfields were once again out of service. The 307th alone had destroyed about a hundred MiG-15s on the ground. One more airfield remained on the target list: Saamcham.

Jack was the tail gunner for B-29 serial number 44-61748, one of the few B-29s in the 307th that had nose art, introducing herself as *Hawg Wild*. Strategic Air Command regulations strictly prohibited such unique identifiers on its aircraft intended to penetrate Soviet airspace to deliver atomic bombs. The 307th arrived at Kadena in the summer of 1950 with no lewd names or racy paintings. Its companion wing at Kadena, the 19th, was not assigned to SAC so it's B-29s wore all the trappings of the glory the bombers had earned in the Pacific War, including colorful, sometimes vulgar, nose art.

---

247. Far Eastern Air Forces Bomber Command, "Operations and Intelligence Summary," June 1953, AFHRA Reel K7181, 851.
248. Robert F. Futrell, *The United States Air Force in Korea 1950-1953*, Revised Edition, (Washington, DC: Office of Air Force History United States Air Force, 1983), 681.

No one knew the story of how *Hawg Wild* got its name. It was delivered to the Army Air Force a few days after the end of World War II in the Pacific and sat idle in the desert storage base at arid Davis-Monthan Field outside of Tucson, Arizona. As Strategic Air Command expanded during the early Cold War and attrition took its toll in the Korean War, many B-29s were brought out of storage, refurbished, and moved to U.S. air bases for newly formed SAC wings or flown to Kadena and Japan as replacement aircraft for the war. Tail Number 748 arrived at Kadena with its nose art on full display in vivid violation of SAC orders.

Jack was ready for the 307th Bomb Group's 573rd mission. He had a full and sufficient breakfast, and a buzz circulated among the combat crews that the war was about to end for sure. Armed Forces Radio provided soldiers, sailors, and airmen across the theater of war with daily updates on the progress of the cease-fire talks at Panmunjom. Negotiators had settled the big issues, and the documents were in the hands of staff officers to hammer out final details and draw the demilitarized zone on the map. Maybe this mission would be the war's last.

Still, it was a combat mission. Enemy airspace was filled with more antiaircraft artillery and searchlights than at any previous time in the war. MiG-15s were present in greater numbers than ever before in MiG Alley. Chinese and North Korean pilots were just about as good as their Soviet trainers. Air battle was still about closing with and killing the enemy before he could do the same to you. The atmosphere in the briefing room for mission 573 was filled with the same smoke-filled, zippo-lighter-flicking, nervous-chatter environment as was Mission Number 1.

"Group, attenshun!" bellowed the adjutant as Commanding Officer Colonel Austin J. Russell strode across the front platform, followed by staff briefers. "Take seats!" as the colonel sat in the front row and the briefers lined up in order of their sequence to take the podium.

By custom, the combat crews sat in groups by position—pilots, navigators, bombardiers, and so forth: the gunners were in the rear. The briefing included information about the target, fighter support, details about the bomb run, and what type of enemy air and ground fire to expect. The group's first B-29 was scheduled to take off at 1900 hours followed by the remaining planes at one-minute intervals. The target, Saamcham Airfield, was just inside MiG Alley well to the north in North Korea. Jack was in for a long and risky weekend.

The briefing offered no indication that this might be the last mission flown in the Korean War by the B-29 Medium Bomber. The dozen crews scheduled for the mission had no hint that they would be making history in the pre-midnight bomb drop. The objective was to complete the destruction of the last enemy airfield in MiG Alley and however many MiG-15's might be located there.

After the briefing, Jack and the rest of his air crew headed for the ramp where the ground crew assigned to *Hawg Wild* were – for what would be their last time – preparing her for one more combat mission. He helped load 38 500-pound bombs, two photo flash bombs, and ammunition for the 50 caliber machine guns. The photo flash bombs would allow them to take pictures before and after bomb impact. After completing the preflight inspection, it was time for a short rest in the squadron area.[249] While Jack was on the hardstand making his final inspections, word spread along the flight line that the long-awaited truce might be signed in the morning, roughly 12 hours after they would unload their 10 tons of high explosives.

About 1700 hours, Jack reported back to *Hawg Wild* for crew inspection by the aircraft commander and completed the final preflight inspection before boarding. After engine start up, 748 had a mechanical problem that needed to be fixed prior to takeoff. The delay caused *Hawg Wild* to miss its scheduled takeoff position, but the ground crew

---

249. Jack Bernaciak to the author, November 4, 2024.

and the flight engineer worked feverishly to fix it. Thus 748 ended up as the last plane of the mission, taking off at about 2000 hours.[250]

Behind 11 B-29s with 176 propeller blades churning, the tempo of the roar of 44 engines increased. They picked up speed. Then they were airborne. The huge medium bombers of World War II fame, now allegedly "obsolete," started their climb. It was 6 p.m., and in a few aircraft radios were tuned to the Armed Forces Radio station for the regular evening news. It was true! The armistice could be signed in the morning, 20 hours later. The planes droned on, their engines powerfully pulling them higher and higher. The coast of Korea appeared and receded below as the aircraft passed over tall mountains and lush valleys.[251]

The sun set, and a brilliant moon appeared. Suddenly, the sky turned totally dark. The full moon, which had been glowing brilliantly in the night sky, just disappeared. It was a total eclipse. It passed in a few minutes, but Jack saw nothing but darkness from the tail gunner position of the last plane in the formation. The aircraft crossed the battle line. It didn't look like a truce down there. Flashes of artillery were visible.[252]

"We maintained our last place throughout the mission and started our approach to the target about 0200 hours on the 26th," Jack said. "After we started our bomb run, there would be no deviation from the glide path to avoid enemy air or anti-aircraft ground fire. We were using the SHORAN system for the bomb run which was a precise radar method of bombing."[253]

The aircraft turned toward their target. For the pilots, bombardier,

---

250. Jack Bernaciak to the author, November 4, 2024.
251. 307th Bombardment Wing Unit History, July 1953, AFHRA Reel Number N0266, 11.
252. 307th Bombardment Wing Unit History, July 1953, AFHRA Reel Number N0266, 11.
253. Jack Bernaciak to the author, November 4, 2024.

and navigator on the flight deck in the front the aircraft, it seemed something like a truce up there. No searchlights sent their probing fingers of light into the sky. Off in the far distance, a few faint puffs of black smoke appeared; but there was no flak in the vicinity of the Okinawa-based aircraft. No communist fighters appeared – a towering thunderstorm and huge banks of clouds over the Yalu, far to the west, apparently prevented them from leaving their Manchurian sanctuary.[254]

But Jack, looking out the rear of the last aircraft in the line of B-29s, had a panoramic view of the enemy's attempts to catch up with the 307th.

"During our approach to the target," said Jack in an interview, "we encountered enemy searchlights and anti-aircraft ground fire that were guided by radar. Our radio operator had counter measures for the ground radar and was able to jam the searchlights and anti-aircraft fire, and we were able to complete our bomb run without being hit."[255]

Unopposed, the Superforts droned on at bombing altitude. As they approached the target, the bomb bay doors opened. The bombers dropped their bombs and turned homeward. The last B-29 bombing mission of the Korean War was completed.[256]

The return flight got them back to Kadena about 0900 hours July 26, 1953. They headed for debriefing with the intelligence officer and later to the complimentary medicinal whiskey and beer. During debriefing, they found out that an armistice was about to be signed and theirs was the last combat mission of the Korean War. The armistice was signed effective July 27, 1953."[257]

---

254. 307th Bombardment Wing Unit History, July 1953, AFHRA Reel Number N0266, 11.
255. Jack Bernaciak to the author, November 4, 2024.
256. 307th Bombardment Wing Unit History, July 1953, AFHRA Reel Number N0266, 11.
257. Jack Bernaciak to the author, November 4, 2024.

A2C Jack Bernaciak had been in the rearmost position of the last B-29 to fly a bombing mission over North Korea during the Korean War. That gave him the distinction of being the last man among bomber crews to leave hostile airspace during the war.

The Air Force retired *Hawg Wild* to the Navy's China Lake Naval Weapons Center to serve as a ground target for crews of "top gun" pilots. It was restored to flyable condition by Jack Kern and his company Aeroservices, of Tucson, Arizona, beginning in 1978,[258] and flown across the Atlantic to the Imperial War Museum's American Air Museum in Duxford, England, in 1980.[259]

Jack Bernaciak built up a successful accounting practice in San Francisco and in 2019 retired to the good life.

Airman 2nd Class Bernaciak in combat crew training at Randolph Air Force Base, Texas. (Photo courtesy Jack Bernaciak)

---

258. "A Trip Across the Pond," Vintage Aircraft http://www.twinbeech.com/B-29HawgWild.htm ; "Old Hawg New Tricks, Preservation History Duxford B-29 *Aeroplane* Issue 563, March 2020. https://www.docdroid.net/2PK9ajD/preservation-history-duxford-b-29-aeroplane-issue-563-march-2020-pdf
259. Mike Overstreet, Director, *B-29 Bomber, It's Hawg Wild* (Tucson, AZ: Key Productions, http://keyproductionstv.com/video/hawgwild/index.html

Jack Bernaciak's combat crew in training at Randolph Air Force Base, Texas. Front Row: Captain Richard T (Ted) Heslam, Radar Operator; Major John P Boukus, Pilot (right seat); Lt Col James M Crick, Aircraft Commander; Name unknown, Navigator; Lt Paul L O'Donnell, Bombardier. Back Row: A/1C Leonard J (Jack) Bernaciak, Tail Gunner; Staff Sgt William Sutherland, Flight Engineer; A/1C Kenneth R (Ken) Hoeffner, Right Gunner; A/1C Phillip R (Phil) MacKenzie, Left Gunner; Name unknown, Radio Operator; A/1C Paul M Savko, Central Fire Control Gunner. (Photo and caption courtesy Jack Bernaciak)

Bernaciak's crew wins "Lead Crew" designation at Randolph Air Force Base before deploying to Korea. Left to right: 1st Lt John W Lewis, Navigator; Captain Richard (Ted) Heslam, Radar Operator; Captain Paul L O'Donnell, Bombardier; Major John P Bokus, Pilot (right seat); Lt Col James M Crick, Aircraft Commander; Brig Gen J H Davies; A/2C Paul M Savko, Central Fire Control Gunner; A/2C Phillip R MacKenzie, Left Gunner; A/2c Leonard (Jack) Bernaciak, Tail Gunner; Staff Sgt William (Bill) Sutherland, Flight Engineer; Name Unknown, Radio Operator. (Photo and caption courtesy Jack Bernaciak)

A2C Bernaciak servicing his tail gun turret during combat crew training at Randolph Air Force Base, Texas. (Photo courtesy Jack Bernaciak)

Airman 1st Class Leonard (Jack) Bernaciak receives the Air Medal from his Commanding Officer on completion of his final mission in the Korean War. (Photo courtesy Jack Bernaciak)

"Hawg Wild" at rest at China Lake as a target for aerial gunners. (https://www.docdroid.net/2PK9ajD/preservation-history-duxford-b-29-aeroplane-issue-563-march-2020-pdf)

"Hawg Wild" on static display at the American Air Museum of the Imperial War Museum Duxford, England. (https://www.airmuseumsuk. org/museum/IWMDuxford/AmHall/800/images/018%20Boeing%20 TB-29%20Superfortress%20-%20Hawg%20Wild.jpg)

Jack Bernaciak's restored tail gun section of B-19 "Hawg Wild" on display at the Imperial War Museum at Duxbury, England. (https://www. airmuseumsuk.org/museum/IWMDuxford/AmHall/800/images/018%20 Boeing%20TB-29%20Superfortress%20-%20Hawg%20Wild.jpg)

On occasion, the Imperial War Museum rolls "Hawg Wild" out onto the ramp for outdoor viewing. (https://www.airmuseumsuk.org/museum/ IWMDuxford/AmHall/800/images/018%20Boeing%20TB-29%20 Superfortress%20-%20Hawg%20Wild.jpg)

Jack Bernaciak (left) and Romaine Gregg say good-bye following the closing banquet of the 307th Bomb Wing/Group (1946-1954) Reunion April 28, 2012. The 16th reunion, held in Orlando, Fla., was attended by former ground crew and flight crew members of the 307th and their families. (U.S. Air Force photo/Master Sgt. Mary Hinson. https://www.307bw.afrc.af.mil/News/Art/igphoto/2000153211/)

CHAPTER EIGHT

Who Were These Men?

Pundits and analysts have said the B-29 was obsolete, outclassed, and too slow for the Korean War. They are wrong. Totally wrong. Really. The B-29's Korean War gunners offer eyewitness accounts to prove the point.

Over the course of the Korean War, MiG-15s shot down 16 B-29s over North Korea.[260] Official records credit B-29 gunners with 25 MiG-15 Kills.[261] Three of those gunners shot down two MiGs each. And one B-29, "Command Decision," became an aircraft ace by shooting down five MiG-15s with different combat crews. That's the post-war record of officially credited enemy aircraft destruction by B-29 gunners.

Even when shot down, B-29 gunners continued to fight the fight against starvation, indoctrination, deception, interrogation, and torture perpetrated by their communist captors. In addition to Philip Aaronson, 307th Gunners Dan Oldewage and Marvin King carried on the fight in POW camps, as told in the remarkable memoir by Marvin King's son, *The Yalu River Boys*.[262]

---

260. Korean War Project, Korean War Air Loss Database KORWALD 15 March 2015. https://www.koreanwar.org/dpaa/korwald-all.pdf
261. Far East Air Forces Official Credit for Destruction of Aircraft 1951-1953, Air Force Historical Research Agency, Reel K1108, pp. 396-670.
262. Dan King, *The Yalu River Boys: The True Story of a B-29 Bomber Crew's Combat and Captivity in the Korean War*, (North Charleston, SC: CreateSpace & Pacific Press, 2018).

The gunners of the Korean War were apex predators. The ten Airmen highlighted in this book of the Korean War are giants among men alongside 307th Bombardment Group World War II gunners I have met, such as Earle McGuire and Dale Strickrath. There are hundreds of others.

They came from diverse family backgrounds: urban or rural; Christian, Jewish, or secular; well-off and poor; from close-knit or estranged families. Some were high performers in school, others nearly flunked out. They defy demographic stratification.

The Korean War experience forged them in fire to create a lasting brotherhood. Each of the three Bombardment Wings that fought the duration of the Korean War have associations dedicated to sustaining the memory, comradery, and legacy of their wartime experiences. The Korean War Veterans of America has over 15,000 members in 129 chapters across 41 states. In each of these organizations, gunners hold a special place in the hearts and minds of Korean War veterans.

So, who were these men? What was it about growing up in 1930s and 1940s America that produced such warriors?

I am a military analyst, not a social psychologist. I cannot provide empirical data to test hypotheses formally. But after examining tens of thousands of pages of archival material, reading hundreds of memoirs and personal accounts of wartime experiences, meeting personally with dozens of Korean War veterans, and studying the experiences of B-29 gunners, I can offer some observations about the character of these brave warriors.

Of course, the gunner had to be adept at assembling and disassembling the weapon. The M3 needs special attention to detail to ensure its headspace and timing are always set precisely. Headspace is the distance between the rear of the barrel and the face of the bolt. Timing is the adjustment of the gun so that it fires when the recoiling parts are in the correct position. On the B-29, unlike the ground-mounted version of the Browning M2, the gunner cannot access the guns themselves when in flight. So, these settings must not only be correct on take-off, but they also need to be perfect after firing hundreds of rounds in flight. The gunner had to apply astute judgement to adapt the settings to account for the number of rounds that would be slammed through the feeder, bolt, and extractor in the course of a combat mission. Set them too close, and the gun will jam during an engagement. Set them too far, and they will not fire on the first press of the electrical trigger.

The gunner had to be mechanically inclined. For a B-29 gunner in the Korean War, it went beyond manual dexterity and handiness with tools. Machine guns require attention to details with thousands of small parts fit together in perfect intricacy to deliver lethal projectiles. Keeping the M3 caliber .50 machine guns ready to fire for a few seconds when it matters most meant gunners had to devote hours to maintaining their equipment each and every day, checking all parts for wear and damage.

The gunner could take no shortcuts in cleaning and lubrication. He had to attend to these tedious details daily, often hourly. Sometimes in the heat of battle, he had to correct a stoppage caused by explosive powder or metal-on-metal contact of parts being slammed together under immense pressure and recoil. The M3 caliber .50 cannot be left idle when not being fired. Especially in the salty, humid atmosphere of a tropical Pacific Island, corrosion and rust spread on metal parts like a virus run amok in a human pulmonary system. Diligence must win the battle with tedium in the heart of a B-29 gunner.

Being a B-29 gunner also required the expertise of an electrician to deal with the complex wiring of the bomber's Remote Control Turret system. The gunner had to know which wire connected to what plug and keep a spool of electrical tape to splice things back together when a circuit went down, even as MiG-15s were attacking the bomber.

All this adds up to a requirement for a gunner to be *diligent*.

The Korean War gunner also had to be *disciplined*.

Just as with the World War II gunner, the B-29 Korean War gunner had to have exquisite hand-brain-eye coordination. Thus, basketball players who instinctively pass the ball to the shooter with the hot hand and baseball players who know how to run to where a fly ball was going to became excellent gunners. They had honed the skill of connecting the eyes through the brain to the hands to get bullets to land where the target was going to be a few seconds ahead of where they were at the moment of shooting. Cognitive scientists today call this phenomenon "heuristics."[263]

B-29 gunners had to internalize their battle drills so they could do their duties without referring to a checklist or technical manual. When seconds count and written instructions were minutes away, gunners of any intelligence quotient had no time to reach for a reference or consult a manual. This applied not only to mid-air engagements. It also was necessary for take-off, landing, and in-flight procedures. The gunner's duties were eclectic, ranging from the machine guns and turrets to engine performance, flaps, and landing gear, as well as external hazards.

When you look at the photos of the gunners in this book, you can sometimes see that gleam in the eyes that can betray the rascal inside. Yet, those who may have been a hellraiser in the full vigor of youth rarely required harsh discipline from the chain-of-command.

To be sure, these men who confronted their enemy 20,000 feet above hostile ground felt the fear that comes with a close encounter with their mortality. But B-29 gunners developed the fortitude that led them to heed Stonewall Jackson's maxim never to give counsel to their fears. Their faith in themselves, their crew, their aircraft, and their guns overcame the alarm, panic, and distress that too often can freeze a human into inaction. Their faith in a God who is real calmed their

---

263. Gerd Gigerenzer, Peter M. Todd, and the ABC Research Group, *Simple Heuristics That Make Us Smart* (New York, NY: Oxford University Press, 1999).

anxieties, allowing them to use their natural abilities and learned skills to fight rather than flee.

The Air Force did away with gunners, and gunner's wings, in 1955, as jet bombers replaced the propeller driven aircraft. Their ranks are thinning as the "Silent Generation,"[264] those who went through the Great Depression and World War II then fought in the Korean War, ages. But their legacy lives on in the aircrew who today defend Air Force bombers such as the B-52, B-1, B-2, and the soon-to-be operational B-21 with electrons and photons in place of bullets.

Finally, I found these men to be *devout*. They were devoted to family, faith, and country. In every oral history and memoir of these men at war, each invariably began his story with a description of their family situation before deploying to the Korean Peninsula. Some were from broken families, men who sought and finally found a father figure or brotherhood among his crewmates. Others came from tightly knit families who found surrogate parents and siblings with whom they bonded for life.

Many openly declared their faith in God while in combat. Some acknowledged after the war that there were no non-believers in a B-29, just as there were no atheists in foxholes. There is something about close encounters with mortality that brings men closer to the path to eternal life.

While they might not often have openly confessed their love of country at the time, it seems that once they raised their right hand and swore "to support and defend the Constitution of the United States of America against all enemies foreign and domestic…and bear true faith and allegiance to the same," their love of country became a glowing ember ever ready to be fanned into flame. From that moment on, that flame became an inextinguishable fire forging the experience of

---

264. "People: THE YOUNGER GENERATION," *Time*, November 5, 1951. https://content.time.com/time/subscriber/article/0,33009,856950-1,00.html

war with all of life's future encounters. I have witnessed the tear-filled eyes of these giants among men as they honor the passing of comrades into eternal life. Their numbers are inevitably shrinking as their generation passes the torch to the next cohort of soldiers, sailors, airmen, and Marines who have taken on the burden of defending freedom.

We must carry on with their mission as they press on to the mark of their high calling.

Afterword

In a world where history is often relegated to the back pages of textbooks and the quiet corners of collective memory, *Gunners!* emerges not only as a tribute to the men of the Korean War Air Force, but as a vessel of healing—an opportunity to give voice to a generation that has so often carried its burdens in silence. For the Korean War veterans among us—now few in number, but mighty in spirit—this book offers not just a reflection, but a release. For their families and friends, it provides a bridge to understanding, empathy, and gratitude.

These Airmen, whose stories fill these pages, lived through bitter cold, long missions, haunting uncertainty, and the deep scars of combat. They returned home without parades or fanfare, and, true to the character of their generation, they quietly resumed their lives—raised families, built careers, and bore their memories with dignity. Yet many of them carried within them the hidden weight of war: trauma left unspoken, experiences too painful or complex to share in everyday conversation.

Gunners! is more than a chronicle of valor—it is an invitation. For veterans, it invites reflection and recollection. It names experiences that may have long remained unnamed. It gently opens a door to memory, allowing old stories to surface—some with pride, others with pain, but all with meaning. In the clinical world, we understand that storytelling—particularly in the context of shared experience—is one of the most effective tools for post-traumatic processing. It allows veterans to reframe their own pasts, not as isolated events, but as part of a greater whole, shared with others who lived through the same crucible.

Reading these stories can prompt a powerful kind of remembrance. Veterans may find themselves saying, "That was me," or "That's what I went through," or even for the first time, "I didn't know anyone else felt that way." And in doing so, they may also find a measure of peace—a reclaiming of meaning, identity, and recognition. Families, too, may discover a new language for understanding what their fathers, grandfathers, or great-grandfathers have endured. Through the lens of *Gunners!*, they can listen not only to the words on the page, but to the conversations these stories ignite around kitchen tables, in living rooms, and during long-awaited moments of vulnerability.

This is why *Gunners!* matters so deeply. Not only as a historical document or a military account, but as a therapeutic instrument of remembrance and reconciliation. These men—patriots all—faced trauma in the air and returned home to face silence on the ground. This book breaks that silence with honor and care.

The ten Airmen profiled in these pages were ordinary men thrust into extraordinary circumstances. Their courage, endurance, and quiet patriotism stand as examples of the strength of the human spirit. But their stories also reflect the very real psychological cost of war, and in doing so, offer a chance—perhaps the last—for surviving veterans to speak what has gone unspoken, to feel what has long been buried, and to heal in the presence of understanding.

I believe that *Gunners!* will do more than educate. It will connect generations. It will prompt conversation. It will bring laughter and tears. And it will, for some, be the first step in putting decades of silence into words.

I urge readers—veterans, family members, and the curious

alike—not only to read this book, but to sit with it, reflect on it, and talk about it. Especially with the Korean War veterans still among us. For in those conversations, we not only honor the past—we heal it.

Tom W. Ayala, PhD
Clinical Psychologist and Professional Counsellor

Acknowledgments

Three of these gunners were alive at the time I was compiling this book. Jack Bernaciak, Romaine Gregg, and Tom Stevens were generous in giving of their time, recollections and photographs to bring their stories to life.

The 307th Bombardment Group Association received me into their membership, providing support and encouragement. Current President Jim Walsh leads this dedicated group by example. Previous President John Poggi connected me with his fellow Vietnam War veteran, Bob Babcock, and Deeds Publishing.

Several US Government workers provided me with generous guidance and support in finding and collecting archival documents, especially Tammy Horton of the Air Force Historical Research Agency and Richard Salyer from the National Archives and Research Administration. The 307th Bombardment Wing remains on vigilant duty in defense of the nation today. Their official Historian, Amy Russell, has been energetically supportive of my research for several years.

I have called on many nonprofit organizations to learn about and meet Korean War Veterans. The Shenandoah Valley Chapter 313 of the Korean War Veterans Association opened their arms to my participation in their work, enabling me to get to know personally several dozen living Korean War veterans. Commander Doug Hall and Previous Commander Narce Caliva offered organizational support, and guidance on contacting Korean War Veterans for this project.

Several contributors are deserving of mention, including my son Josh Blackwell, a commercial pilot who corrected many of my clumsy attempts to describe how airplanes fly; my son Jonathan, an Air Traffic Controller whose meticulous attention to detail and creative mind helped guide my research and composition; and my daughter Elizabeth Baldridge, whose connections to the real world as a public school educator have given me sound advice on how to descend from the heights of my academic writing to produce a book that the broader public might find appealing. Our Jewish next-door neighbors, Harvey and Sandy Ascher, whose own family POW experience gave them deep insight on what the Korean War POW experience was like for those of Hebrew legacy, provided me with unique insight into what it was like to be an Jewish American Airman in the 1950s.

Eric Minton served as my professional editor for *Gunners!*. Eric is an award-winning photojournalist whose fulltime day job is Editor for investigative bodies of the United States Government. His ruthless red pen and sharp creative mind have always served me well. My brother Jeff Blackwell has ever been my harshest constructive critic.

This is my third major publication. Through each, and especially this one, my wife Rosalie Marie (LaFont) Blackwell has inspired and supported my efforts. Her father was a combat wounded veteran of World War II who marched with the infantry from the Normandy Campaign and across the entire West European continent to Victory in Europe Day. Her Cajun spirit and Christian faith remain my inspiration.

www.ingramcontent.com/pod-product-compliance
Lightning Source LLC
Chambersburg PA
CBHW051606170426
43196CB00038B/2946